Artificial Gemstones

Dedicated to the memory of Tony French,
and to Nola and Raoul

MICHAEL O'DONOGHUE

Artificial Gemstones

N.A.G. PRESS
an imprint of
ROBERT HALE · LONDON

© *Michael O'Donoghue 2005*
First published in Great Britain 2005

ISBN 0 7198 0311 X

NAG Press
Clerkenwell House
Clerkenwell Green
London EC1R 0HT

NAG Press is an imprint of Robert Hale Limited

The right of Michael O'Donoghue to be identified as
author of this work has been asserted by him
in accordance with the Copyright, Designs and
Patents Act 1988.

A catalogue record for this book is available from the British Library

2 4 6 8 10 9 7 5 3 1

Typeset by e-type, Liverpool
Printed by Kyodo Printing Co. (S'pore) Pte Limited

Contents

Acknowledgements

The author and publishers wish to thank most sincerely the late Professor Dr Eduard Gübelin for permission to reproduce photographs from his collection and for supplying captions to them.

We also wish to thank Alan Jobbins, Kenneth Scarratt and Gem-A for the use of their photographs.

Most grateful thanks are also due to Lorne Stather, FGA, DGA, for a critical reading of the manuscript. Lorne's experience was invaluable and cheerfully given. I am, as always, grateful for Louise Joyner's support and encouragement.

Illustrations

Black & white figures

Illustration Credits

Colour

Dr Edward Gübelin: 1–28, 30–46, 53–6 & 63–75.
GemA: 29, 47–52 & 57–62.

Black and white figures

Reproduced from *Gems Made by Man* with permission from Kurt Nassau.

Introduction

Y ears of university teaching have convinced me that in many minds each new synthetic material and each new treatment or imitation somehow sweeps older examples out of sight. Readers should remember that synthetic rutile masquerading as diamond and garnet-topped doublets are still around.

Readers might also be forgiven for thinking that contemporary gemmological techniques (especially Raman spectroscopy) have taken care of simple synthetics and imitations once and for all. In one sense this is true, since apparatus now available can detect virtually anything: such machines, however, are not available to all and their operation and the interpretation of the results obtained need a degree of specialism not provided for in relatively simple gemmology courses.

So in one sense gemmologists are still where they were in 1983 – they may still be confused but at a higher level. The aim of this new book is to introduce readers to the latest man-made products and their detection, to the techniques and assessment of colour enhancement and to act as a museum showcase for products that are still manufactured and worn after well over a century.

We shall review crystal growth techniques with emphasis on those aspects of a capricious process which affect gemmologists most. No more technical details will be given than are necessary for a reasonable and practical understanding of product identification. Details of the growth of gem-quality diamond are given only in general terms since much of the process does not affect those tests which the gemmologist can use. In any case the processes are not widely published, for obvious commercial reasons.

I have arranged the text in such a way that the major gem species are covered first. Within each chapter the main description precedes notes of recent or significant developments taken from current literature. This follows the practice of the first edition.

We are considering artificial gemstones, so less attention is paid to organic materials as they are not synthesized but only imitated. The various methods of growing cultured pearl are not described: these materials, both natural and artificial, are well covered in a recent book by Maggie Campbell-Pedersen, *Gem and Ornamental Materials of Organic Origin* (Butterworth-Heinemann).

Some of the history of artificial gemstones can be followed, by anyone wishing to spend the time, in the serious and – more often – the popular press. For a body which prides itself on a sense of adventure and risk-taking, the gemstone and diamond trade has often been represented as the victim of scares and shocks: once it was synthetic ruby, then the cultured pearl, synthetic emerald, YAG, CZ, moissanite and now synthetic diamond. I have seen no signs of the trade collapsing with worry over the appearance of any of these materials so it is odd that it should be so represented. In any case, education can, should and does help a great deal towards the avoidance of catastrophic mistakes.

Most countries with important gem deposits and those which have a significant jewellery trade have some sort of gemmological institution which instructs and examines in gemstone identification, now known universally as gemmology, and which validates the qualifications obtained by success in the examinations which are based on practical identification of unknown specimens. The examinations are searching – what use would they otherwise be? – and also include questions on identification techniques, to be answered in writing.

This seems the appropriate place to explain that the terms natural, synthetic and imitation have specific meanings as far as gemstones are concerned: a natural specimen occurs by the operation of natural processes. All synthetic stones have a natural counterpart with which they share identical chemical and physical composition (with the occasional small variation); an imitation stone is anything which looks like one species while being a member of another. In this way one natural gemstone can imitate another and should a collector of synthetic gemstones wish for imitations they can be natural ones. This seems pretty obvious but some of the gemmological journals still run this topic by their readers, who should feel indignant that space and time are being wasted in this way. My use of the adjective 'artificial' embraces all products that have not arisen naturally.

Collectors of synthetic gemstones do exist and the range of products is far wider than might be imagined; while many man-made substances may not be suitable for wear, their crystals can be very beautiful indeed and occasionally grace gem and crystal shows, so that I have felt it worthwhile to mention them here.

Readers wishing to further their knowledge of artificial gemstones should consult the Bibliography. Serious students (is there another kind?) need access to national or the largest university libraries, which may carry the range of journals that is needed. Access is not always easy but should not be impossible providing borrowing rights are not required – these are usually harder to secure. In any case much of the necessary data are now available on the Internet.

It may seem odd to boost journals at the expense of monographs in a monograph but the fact remains that much of the data needed in a study of artificial gemstones are numerical and constantly in danger of being superseded. Monographs appear at longer intervals but provide scope for longer discussions of both numerical and non-numerical topics. They can also carry photographs – journal editors have restrictions on the number of photographs they can accommodate. Monographs are far more likely to be found in libraries than runs of journals.

One or two specialist libraries exist but staff numbers are generally small. It is essential and sensible to write to make an appointment to view the collections as a whole before commencing serious study: it may very well be that the library of choice will turn out not to hold the material you want. Remember that curatorial staff, at least in the national libraries, are likely to be specialists in their fields and although all enquiries are certain to be received courteously, elementary problems could well be solved by recourse to encyclopedias and textbooks.

The Internet is an excellent source for all kinds of information; the choice of service provider and search engine is a matter of personal preference, but many libraries will give advice on searching techniques.

As I have already noted, the book not only deals with gem materials made by man but also gives instances of the now very common practice of gemstone enhancement. This is the improvement of a stone's colour to make it more attractive to the public and the trade. With the coming of synthetic diamond, colour enhancement has made the life of gemmologists and those working in laboratories much more complicated and now that customers demand that all but the least significant stones are sold and resold with an accompanying certificate great care is needed before anyone can say with confidence that a particular stone is natural or synthetic, has a natural colour or one which has been altered or improved.

It was customary at one time for gemstone surveys to include lists of trade names though hardly any were ever used. The number of names given to ornamental materials by different trades involved in handling them is only equalled, if not surpassed, by the names given over the centuries by mineralogists. I have given a list of some of these names and have mentioned, in passing, those which genuinely appear to be used today and whose absence might pose problems.

In recent years there have been well-worked attempts to produce lists of mineral names and some artificial gemstone names are included. The most useful of these name lists at the time of writing is Peter Bayliss, *Glossary of Obsolete Mineral Names*. Synthetic materials are included as well as such old friends as Swiss lapis (dyed jasper).

The student of artificial gemstones will sooner or later need to consult authorities in which up-to-date details of minerals species are given. For many years there was only one single source for this information: M.H. Hey's *An Index of Mineral Species and Varieties Arranged Chemically*. Readers can look up the species name in the index then consult the main text in which it will be found among its chemical relatives. In many cases artificial substances and some trade names are included.

The names of some artificial materials are also given in *A Manual of New Mineral Names*, by Peter Embrey and John Fuller.

In Andrew Clark's *Hey's Mineral Index*, mineral species, varieties and synonyms are arranged alphabetically, and it also includes the names given to artificial minerals. The best reference in English is provided where appropriate.

It is not always possible to identify an artificial gemstone with the relatively simple instruments available to the gemmologist, but instruments are graphically described by Peter Read in *Gemmological Instruments*. Though some models and addresses of suppliers will have changed since then this is still an excellent guide and should be obtained by anyone who is likely to be concerned with making the distinction between natural and artificial gemstones.

CHAPTER ONE

The Structure and Development of Crystals

The majority of artificial gemstones are crystalline. No one should be surprised at this since their natural counterparts, where they exist, are crystalline too. To understand some of the methods by which specimens of either kind can be identified, some facts on the crystalline state are necessary.

A crystal is commonly defined as a naturally occurring solid whose component atoms display long-range arrangement. The atoms which make up gases and liquids do not display such an order, being in continuous rapid movement. Solids in which the atoms show no kind of order are rare in nature: glass and opal, the examples described in this study, are virtually the only amorphous (non-crystalline) solid materials known. In these substances the constituent atoms show only short-range order.

CRYSTAL MORPHOLOGY

The regularity of the atomic arrangements shown by crystals is responsible for the array of faces and edges seen in many specimens, although not all crystalline materials invariably show such features all the time – circumstances of growth see to that. Crystals grown in the laboratory often show a fine display of faces, if they are grown in certain ways. Such crystals, in fact, may well be collectors' items for this reason; good examples include the magnificent rubies grown by the late Professor Paul Otto Knischka in Austria.

We can all see that the arrangements of faces and edges give particular recognizable shapes to some crystals, but the language of description needs some comment. With the Internet it is easy to send a picture of a crystal to anyone but for strict description the terms based on the visible structures given by the internal arrangements need to be used. In this text the descriptions and terms are simplified.

The word 'form' has a number of possible meanings in English but when we are describing crystals it means the particular appearance (faces and edges) of a crystal which is the result of the underlying atomic structure. We are using a 'form noun' when we use the word 'cube', a shape in which all the angles are 90 degrees and all the dimensions are equal to one another. This is the simplest form to understand and describe and we can conveniently remember that when discussing a crystal we are thinking in three-dimensional symmetry (which makes it hard when we have only the two-dimensional page).

Before describing a crystal we have to refer the three-dimensional body to three, sometimes four, imaginary axes which define the crystal's position in space or, more understandably, why the faces are particular shapes and sizes and why they meet one another at particular angles. Because we refer a crystal to these axes they are called axes of reference, or crystallographic axes.

Three-dimensional atomic symmetry cannot give an infinite number of possible shapes (the theoretical total is 230 but some have never been found existing) so the outward forms are limited to relatively few. Circumstances of growth (the slower the growth, the larger the crystal faces) give the final 'habit' of the crystal – loosely speaking habit is form modified by other influences. Some of the nouns and adjectives used to describe simple forms are the same as those used to describe habit: here are some examples:

- prism form, with faces meeting in parallel edges, when predominant in a crystal gives prismatic habit
- pyramidal form, where faces when produced (extended) when necessary will intersect the three axes of reference
- pinacoid form, where prism faces are subordinate to flat faces (pinacoids) at right angles to them; in practice crystals of this type are known as tabular or flat

These three are the most significant forms for the crystals we shall encounter in this survey.

Some common habits include acicular (needle-like and not much use for ornament), prismatic (prism form dominates and crystals are usually longer rather than fatter, with a regular cross-section which can help in identification), botryoidal (resembling a bunch of grapes) and reniform (like a kidney). The last two do not occur in synthetic gemstones.

Crystals in general can be divided into seven crystals systems, again on the basis of their atomic arrangement, but we do not need to study this area here apart from knowing the names of the systems which might creep into some of the descriptions. In descending order of symmetry they are: cubic (or isometric), tetragonal, orthorhombic, monoclinic, triclinic, hexagonal and trigonal (the latter two are referred to four as well as three axes).

Crystals showing a number of right angles may well be cubic and those with none may well be triclinic, a system in which no right angles are possible. To ensure that we are thinking along the same lines, an octahedron (two pyramids of equal dimensions joined base to base) can be formed by cutting down the corners of a cube. This places diamond, of which the octahedron is a favourite form, in the cubic crystal system with octahedral form and habit.

CRYSTAL GROWTH

All these factors have to be borne in mind if one is going to grow a crystal. How one grows a particular substance depends on what it is going to be used for. It is no good producing an acicular crystal if it is going to be fashioned into a gemstone. (Similarly it should not be so light that it will not hang properly in a necklace nor so heavy that as a brooch it pulls away from the clothes.)

The chemical nature of the material to be grown is of great import- ance. I am sometimes asked why there are no synthetic tourmalines or silicate garnets. It is because both these gem minerals are compounds, quite complicated chemically, with several components which will have different melting points. It is easy to melt a substance but cooling it at just that rate which will avoid incongruent melting and conse- quent production of an unsatisfactory solid takes a good deal of prior research as well as engineering skill.

Readers may well think that diamond, being an element not a compound, will not present problems of this kind. This is true but in the growth context, as in so many other respects, diamond is on its own: it is uniquely hard and tough, needing high temperatures and pressures to produce a crystal large and pure enough to be used as a gemstone.

I have outlined in passing some of the problems in crystal growing already but the criteria for growing usable crystals are still more demanding. We can grow crystals from solutions with a crystal-growing set (which is well worthwhile and great fun), but the crystals grown this way (copper sulphate, for example) do not have the hardness, toughness and resistance to unfavourable conditions that a crystal of gem quality has to have. The finished crystals need to be large (compared with most other crystals that are grown today), resistant to abrasion and fracture, have colours that do not easily change, be as free as possible from included matter and achieve shapes (forms and habits) that will make fashioning possible.

These are the criteria for the finished crystal. During growth (by the most commonly used methods) the crystal will also have to melt easily, cool evenly, predictably and controllably (computer control is essential in most cases), not form unwanted compounds with substances met during growth and not form groups of small crystals where one alone is required.

A common misconception among crystal growers is that they are all growing gemstones. Of course there are some, probably fewer than 100 in the entire world, who devote all their efforts to these products but by far the majority are producing 'butter on bread' (the butter is what is wanted and the term is substrates). Such crystals as gallium arsenide and other materials known for short as III-V compounds occupy much more personnel and time than gem crystal growth. Without III-V compounds the world would lack many of the electronic instruments now considered essential for life.

Nor are many crystal growers working on new substances with a view to using them for ornament. Some new transparent materials with sufficient hardness, size and attractiveness (transparency, colour) do appear from to time to time but virtually always as spin-offs from research unrelated to gem use.

When we look at the natural gem minerals which might be considered worth synthesizing, we find that, with the exception of diamond, the best-known examples have chemical compositions that are quite easy to produce; the mineral corundum, with the gem varieties ruby and the different colours of sapphire, and the hard spinel which can also be made in many colours, are both oxides, the easiest substances to be grown quite simply from the melt.

Emerald, as a silicate, is chemically more complicated than the oxides of corundum and spinel, so while synthetic emerald is much more expensive than some of the synthetic rubies on the market, it is still reasonably priced considering the effort that has to be made in its growth.

TECHNIQUES

There are several techniques available for growing crystals. The choice depends on the substance required and the price at which the finished crystal is to be sold. The cost of the finished product may range from a few cents per carat for a crystal grown by the cheaper Verneuil flame-fusion method to hundreds of dollars for rubies or emeralds grown by other methods. The main difference between flame fusion and other methods of growth is the time taken for production of the finished crystal. This can range from a year for a flux-grown emerald or ruby to hours for a Verneuil-grown corundum or spinel.

The flame-fusion Method

Inevitably associated with the name of A.V.L. Verneuil the flame-fusion method of growth was not in fact the first technique used to grow ruby. In the now rare and highly collectable *Synthèse du rubis*, written by E. Frémy and published in Paris in 1891, crystals of ruby show as flat plates rather than the boules so characteristic of flame-fusion growth.

Details of the method of production are shown in a number of crystal growth books. A feed powder is produced, consisting of the ingredients of the material to be grown, including any additional elements to give colour or special effects. Such added elements are called dopants (the best known is chromic oxide) which is added to otherwise colourless corundum to give the red needed for ruby.

This feed powder is allowed or assisted to fall by gravity through oxygen and hydrogen flames, which melt it (the melting point of corundum is about 2,037°C). The molten droplets fall and congregate on a rotating/descending ceramic pedestal, where they cool in the shape of the boule, a necked shape which grows quickly into a rounded crystal. Owing to an instability in the feed material which is transferred to the grown crystal the boules are prone to breaking in the vertical (long direction, which corresponds to the major crystal axis for corundum).

The gas from the flames appears in the finished crystal as well-rounded bubbles with bold edges. These are easy to see, a good deal easier than the other major growth feature, the celebrated curved growth lines and colour banding which, although well illustrated in many textbooks over the years, are in fact not at all easy to see (the lines, when one does see them, can accurately resemble the grooves on a vinyl disc). These growth indicators show the gradual build-up of the growing boule (which is a single crystal) and are not seen in crystals grown by other methods.

It cannot be said too often that the lack of natural solid (i.e. mineral) inclusions is the best way to distinguish an artificial gemstone from its natural counterpart. The Verneuil ruby, for example, shows only the growth features and the gas bubbles, while a natural ruby may contain both solid and characteristic liquid inclusions. While these are absent from the Verneuil product, traces of the growth method used in other techniques can at times look suspiciously natural.

Solid inclusions in ruby are necessary to produce the star effect or asterism. In natural ruby a six-rayed star can sometimes be seen when the stone is cut as a flat-bottomed unfaceted but polished dome – a form known as a cabochon. The cause of the star is the titanium dioxide rutile, a mineral which forms acicular crystals. The crystals can, in the appropriate circumstances, form the rays of the star by orienting them-selves at right angles to the vertical axis which in the cabochon runs at right angles from the base through the top of the dome. The star is best viewed under a single source of light.

For this effect to occur in natural ruby rutile needs to be present in such a way that heating subsequent to formation causes it to move from being disseminated in dust-like particles throughout the crystal to take the star position. In a synthetic ruby rutile is added to the feed powder and the finished boule reheated to induce the rutile to make the rays of the star.

The star effect can also be found in nature in blue sapphire which has, of course, the same composition as ruby, although the blue colour needs the addition of both titanium and iron as dopants. Reproduction of the star effect can take place only in the flame-fusion method of growth of either ruby or blue sapphire. It cannot form in the other colours of sapphire as rutile is not present; for the same reason Thai rubies, which are rutile-poor, do not show the stars seen in Burma or Sri Lanka stones.

In Verneuil star rubies and blue sapphires the rays of the star appear too perfect and the base of the cabochon is usually polished flat. In natural star corundum the base is usually left lumpy, as this means that the stone weighs more and can command a higher price (the dealers charge a certain amount per carat for all but the very cheapest stones).

Spinel, which is also a hard oxide with a simple atomic structure, is also easily grown by flame fusion, although there are some differences in the procedure. To achieve a satisfactory boule of spinel the chemical composition has to be different from that of the natural material. In general to get a boule of colourless or blue spinel the composition has to change from $MgAl_2O_4$ to $Mg2\frac{1}{2}Al_2O_4$.

Even when this modification is made some colours will not grow satisfactorily. Red spinel, the most desirable colour, needs chromium as a dopant and the process seems to be difficult.

Flux Growth

Rubies and emeralds in particular, of the major gem species, are grown by a method in which the starting materials are first dissolved in a compound known as a flux, which then allows the desired substance to grow at a lower temperature. Products grown by this method are expensive. The growth time is months rather than hours; the rate of cooling from the molten state is critical and is regulated by computer. All this costs money, as do the crucibles in which growth takes place: they are made mostly from platinum or iridium, so that unwanted compounds cannot form with the melt of the desired substance.

The advantage of this method is that growth can take place at a lower temperature than would otherwise be needed; moreover, the nutrient can be replenished if necessary since growth takes place in an open crucible.

Materials which enter a different phase at temperatures above their melting point can be grown by this method at a lower temperature, although they cannot be grown simply from a melt. Zirconium oxide (ZrO_2) transforms from cubic to monoclinic at 1,150°C, and this change can be avoided.

Many flux-grown crystals show natural faces, allowing experiments to be made without further polishing. On the other hand crystals are relatively small and ions of the solvent phase enter the crystal lattice. Here a careful choice of flux can minimize any ill effects.

Flux growth for gem materials usually takes place on a prepared seed, although spontaneous nucleation may be allowed for the growth of crystal groups (which happened with the Gilson emerald groups). We have seen that crystals of crucible material, usually platinum for gem crystal growth, turn up as jagged metallic inclusions in faceted stones: it is interesting to note that pure platinum is less likely to succumb to attack by the molten substance than alloys. Nucleation may take place on the melt surface or on the platinum wall; if the crucible is old this may be a preferred site, as the surface will be rougher and encourage nucleation.

METHOD OF GROWTH

The importance of this method makes a full description worthwhile – all the finest synthetic rubies and emeralds are grown by this method as well as a number of other gem materials.

A saturated solution is prepared by keeping the constituents of the desired crystal and the flux at a slightly higher temperature than the saturation temperature for long enough to allow a complete solution to form. The crucible is then cooled through a temperature range in which the desired crystal is known to precipitate. Nucleation usually takes place first in the cooler part of the crucible; there usually *is* a cooler part, although it is possible to keep the crucible isothermal. Once nucleation has taken place no more will occur, so that crystals from the initial sites will be the maximum size. On the completion of cooling, crystals are removed from the crucible by hand or by chemically dissolving the flux.

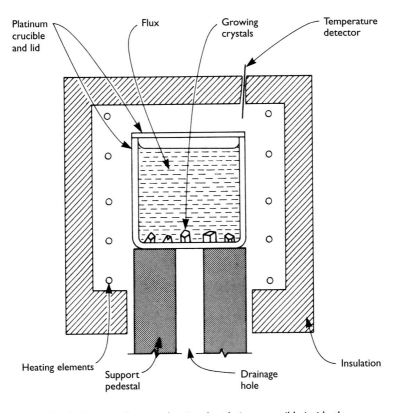

Fig. I Cutaway diagram showing the platinum crucible inside the flux-growth furnace

Growth on a seed gives a greater control of perfection, doping and orientation than spontaneous nucleation. There are two methods in common use:

- growth by slow cooling with a seed
- growth on a seed which is placed in a cooler part of the system while the reset of the solute is in contact with the solvent in a hotter part of the system (growth in a thermal gradient)

The latter is similar to hydrothermal growth but has the advantages of fewer flux inclusions, so that crystal quality and clarity are good. In addition a faster rate of growth can be achieved. The amount of supersaturation can be kept lower than that which is needed to achieve nucleation on the platinum, so that all growth will take place on the seed.

On the other hand, growth with a seed needs more elaborate equipment and more careful control of the growth process. The choice of flux is also more critical. Seeds may have to be specially grown and need to be orientated before growth takes place on them.

The choice of flux depends upon more than one criterion. It must have a low melting point and be a good solvent, dissolving from 5–30 weight per cent of solute at the maximum temperature intended. It must not form a compound with the solute and it should be compatible with the platinum over the intended temperature range. It should have low volatility, if growth is to take place by slow cooling, otherwise a lower temperature range and a faster rate of cooling will be needed. This will often cause nucleation at the melt surface. The flux should have low viscosity and, if possible, be non-toxic. It should also be relatively easy to separate from the crystals at the end of the run.

The advantages of growth by flux evaporation, compared with growth by slow cooling, are that the growth occurs at a higher temperature so that the desired substance does not form compounds with the flux, which is more likely to occur at lower temperatures. If the run is completed, almost 100 per cent of the possible yield is achieved and crystals free from a coating of flux are obtained. Large, well-formed crystals should be obtained from the crucible base, although this will not always happen. On the other hand there is less control over the process and slow cooling will produce better-quality crystals.

YAG has traditionally been grown by this method, one form of growth employing a flux of lead oxide and boron oxide. A mixture of lead oxide, boron oxide, yttrium oxide and aluminium oxide was heated in a platinum crucible for four hours at 1,250°C and cooled at 1°C an hour to 950°C. The crucible was then removed from the furnace and cooled and the solvent leached out with hot nitric acid. Crystals as large as 10x7x7 mm were obtained. Another method used a flux of lead oxide–lead fluoride and boron oxide; crystals up to 100 g were obtained and the optical quality was good.

GROWING EMERALDS BY FLUX GROWTH

In the 1930s emerald was produced by the German firm of IG Farbenindustrie using a flux reaction technique in which pieces of silica were floated on a solution of beryllia and alumina in the correct proportions. Emerald was then precipitated by dissolution and diffusion. Later apparatus separated the reactants from the growth regions by diaphragms. For a flux, alkali vanadates, molybdates and tungstates are suitable; emerald was grown as far back as 1888 using a flux of lithium molybdate and lithium vanadate. A patent granted in the United States in 1967 used the same flux and in this case the emerald was grown on a seed. It is preferable to use materials with a high degree of crystalline perfection and a minimum amount of water as seeds. These are usually high-quality natural beryl, aquamarine or emerald which has been heat treated. Seeds where the faces of the plates are orientated perpendicular to the C-axis are preferred.

A flux of, perhaps, lithium oxide 2.25–3.25 molybdenum oxide is placed in a platinum crucible positioned in the lower section of a vertical tube furnace. The central portion of this furnace is heated to about 1,000°C, subjecting the crucible to a reverse thermal gradient as the hottest part of the melt is at its surface. Chips of beryl or powdered oxide are first added to the melt to adjust the solute-flux composition to an equilibrium state. Then one or more seed plates of natural or synthetic beryl are positioned in the lower, cooler part of the furnace.

A test seed of known weight can be used to assess whether equilibrium growth conditions have been achieved. By frequently removing and weighing a test seed, a point can be found where solution stops and crystallization begins. At this point the seed plates of emerald can be lowered into the melt. It is also possible to keep a test seed plate immersed in the melt and to remove it periodically to assess the growth rate on the other seeds. When emerald crystals of the required size are achieved the crucible is removed from the furnace and the crystals recovered by pouring off most of the flux and dissolving the rest by boiling in an alkali solution.

This method confines growth with beryl structure to the seed and much of the production is optically transparent and substantially flawless. Spontaneous nucleation and twinning on the seed is minimized. Favourable growth rates over a long period can be achieved.

The earliest synthetic emeralds could be identified by the bright red shown through a Chelsea filter; iron content in most natural emeralds prevented them from appearing so bright. In general, although full testing is always needed, the SG and RI tend to be lower than those found in most natural emeralds: SG 2.65 and RI 1.56–1.57 would be typical.

For a while Pierre Gilson added iron to his emeralds, which had the effect of raising the constants and diminishing the fluorescence. Interestingly, I have rarely found emerald to show very much response to LWUV in any case, although the effect is more pronounced with the artificial product.

'Smoke in a still room' is a phrase I have used for years when describing the interior of flux-grown emeralds. The 'smoke' consists of undigested flux, which forms twisted veil-like structures and is the most characteristic sign of this method of growth in any species. The lack of natural solid inclusions, however, is at least as important a clue.

In some flux-grown emeralds squashed liquid drops containing crystals may be seen: some of these may be the beryllium silicate, phenakite. Its presence often suggests some modification of the crystal growth process, sometimes a sudden temperature increase. It has been noted that the phenakite crystals show distinct crystal faces. Many of these effects are best seen under polarized light.

Virtually all synthetic emeralds are transparent to UV rays shorter than about 300 nm. Chatham flux-grown emeralds have been found to transmit freely down to about 230 nm.

While hydrothermal and natural emeralds will show the presence of water when examined by infra-red spectroscopy, this is not seen in the flux-grown product.

OTHER FLUX-GROWN SUBSTANCES

Rubies are also grown by the flux-melt method. The stones are characterized by flux remnants, some of them resembling 'paint splashes', as when a loaded paint brush is slapped against a wall. This effect is especially characteristic of some of the Kashan rubies. Elongated cavities found in flux-grown rubies appear to be grouped and parallel; to the unaided eye they appear whitish. Like flux-grown emeralds, rubies grown in this way show undigested flux in twisted veil-like form; on occasion the undigested flux may take on a roughly hexagonal shape.

Synthetic garnets are more commonly grown by pulling but those crystals grown by the flux method have been found to show liquid-filled channels containing two-phase inclusions. The channels often show a jagged outline and may lead into a network of intermingled feather-like structures rather like those seen in synthetic emeralds.

Another species grown by the flux-melt method is the alexandrite variety of chrysoberyl. The alexandrite grown by crystal pulling and usually intended for research or industrial purposes is too faintly coloured for gem use but the flux-grown variety shows a rather lurid colour change, too strong to be seriously mistaken for the natural material. As far as identification goes the growth method is indicated by twisted veils of flux, minute crystallites and undigested flux particles.

Some colours of spinel have been grown by the flux-melt method. Red spinel which is known not to be natural will very probably have been grown in this way. Botryoidal inclusions or other irregular shapes have been noticed. Colours (other than red) are unlike those of natural spinel and also unlike those shown by Verneuil-grown spinel. Melt residues are reported.

The growth of cubic zirconia (CZ) is difficult, as the material has a melting point of 2,750°C. This rules out the use of any type of crucible. It is grown from a melt of its own powder – the term 'skull-melting' comes from the shape of the apparatus. A block of powder, usually with some pieces of zirconium metal embedded in it, is placed in a cup-shaped framework whose sides are made up of concave and parallel copper tubes which carry circulating cooling water. Heat is supplied by a radio-frequency source (4 MHz and 100 kW). The energy enters the skull between the fingers formed by the copper tubes. The zirconium metal heats first and transmits its heat to the powder, which then conducts the electricity and melts. The metal reacts with oxygen in the air to provide more zirconia. The zirconia melts and leaves only a thin layer next to the copper fingers so that the melt is confined within a container made from its own substance, contamination from the copper being thus avoided.

Since the 1960s a number of hard, transparent materials with no natural counterparts have been grown, largely by flux-melting. These crystals are either left colourless or doped with a variety of substances to give attractive and permanent colours.

Hydrothermal Growth

The term hydrothermal suggests, correctly, that water and heat are necessary for crystal growth. It is, above all, the method chosen for the growth of very large crystals, particularly of quartz, which is so essential for electronic devices. Crystals up to 10 m have been grown in Russia.

Hydrothermal growth takes place in a closed crucible under pressure, which makes for some danger: the crucibles have been called 'bombs' although the correct name is autoclaves. Temperatures may be about 700°C and pressures up to 3,000 atmospheres. They represent an approximation of the conditions obtaining in the Earth's crust when the natural counterparts of the artificially grown crystals were forming.

The autoclaves are sealed and can only be opened and their contents harvested by cutting them open. They are usually made from steel and are lined with precious metal (platinum or gold) to prevent the growth of unwanted compounds; iron, in particular, can be a great nuisance.

The feed material is placed in the bottom of the autoclave, where it dissolves by heat applied beneath. Convection carries the melt upwards in the autoclave where it is allowed to collect on prepared seeds: in the case of quartz the fashioning and orientation of these seeds is especially important when crystals for electronic use are to be grown.

While this method has produced large quantities of colourless quartz (rock crystal), amethyst and citrine, it has been less frequently used for the major gemstones. Emeralds are the most common hydrothermal versions of major gem species. Compared to emeralds produced by flux growth, hydrothermal emeralds may at first sight appear suspiciously inclusion-free and of course contain no natural-seeming solids. There are also no characteristic twisted veils and angular fragments of crucible wall material.

On the other hand hydrothermal emeralds will often show flame-like structures or chevron markings, neither of which is found in natural emerald. These are tapered tubes leading away from the seed in the direction of growth.

Materials which have a high vapour pressure near their melting points, like zincite, are grown successfully in an autoclave. Although zincite is not a well-known gem material, crystals grown in this way can be very attractive (they are doped to produce green or, in particular, a very fine red-orange colour). They are suspected to have masqueraded as natural zincite, a collectible mineral which has sometimes been cut, though in much smaller sizes.

Cheap rubies are easy to make by flame fusion and better quality ones by flux growth, so not many are produced by the more expensive hydrothermal growth. The dopant is chromium, provided by sodium bichromate, and the nutrient either corundum itself or the related mineral gibbsite (aluminium hydroxide). Seed crystals, either synthetic or natural corundum, are suspended from a silver frame in the upper part of the autoclave. Heat to around 400°C is applied to the base of the autoclave.

Hydrothermal emerald is grown in a similar way – fortunately the inclusions can usually be seen after careful examination. The emeralds grown by Lechleitner contain a seed with shallow overgrowth, the seed being cut to the shape of the finished stone. The junction between the seed and the overgrowth is prominently marked by a 'crazy paving' network of cracks.

The synthetic garnet analogue YAG has been made hydrothermally though crystal pulling is preferred. Stoichiometric amounts of corundum and yttrium oxide were placed in platinum capsules with solvents of sodium hydroxide (sometimes others); after a forty-eight-hour growth period, whatever the solvent, YAG resulted.

Hydrothermal growth can take place at lower temperatures than some other methods and thus prevent the development of strain in the crystals. Thermal gradients should be avoided where possible.

CHAPTER TWO

Summary of Treatments

Since the previous edition of this book (1983) gemstones whose colour and sometimes clarity have been altered artificially are more accepted by the gemstone trade. As always, stones treated many years ago are still with us but since the arrival of HP/HT diamonds and synthetic diamonds, as well as beryllium-diffused corundum the trade has had little time to spend on identifying the treatment of less important species.

In the Winter 2002 issue of *Gems & Gemology* Smith and McClure commented on commercially available gem treatments and included a chart on which the treatments were displayed. What follows is based in part on the notes made on the chart and in the commentary.

COLOUR ENHANCEMENT

Dyeing

This form of treatment is appropriate only for stones which are porous enough to take a dye or which have fibres along which dye can travel. Agate, an example of the latter case, is routinely dyed and apart from the rather stronger colours there is no test worth the expense of embarking upon. All agate, so far as the trade is concerned, is dyed.

Dyeing of materials such as turquoise should be detectable with a lens or microscope, as the dye concentrates in surface-reaching cracks or fractures. Often the process is easily spotted if the resulting colours are too lurid to be natural, but recognizing them needs some experience. In the case of banded agate we often find that artificial colouring can be made to look quite like the paler colour that the dye replaces but much more commonly the natural colour is overlaid by too strong a red, brown or black.

Dyeing of jade minerals (almost invariably jadeite) is harder to spot since bleaching may have preceded the dyeing process. While treated green jadeite may sometimes be recognized with a hand spectroscope the other colours (including lavender and yellow), when dyed for enhancement, cannot be recognized in this way. We shall see more about this in the chapter on jade.

Opal is more serious since the play of colour seen against a dark background is more desirable than when the background is light-coloured. The type of opal long known as treated opal matrix shows dark spots of carbon when treated.

The artificial colouration of grey to black or golden pearls needs more advanced testing techniques such as EDXRF, UV, or Raman spectroscopy.

Coral and turquoise, when dyed, are much harder to detect, especially when the colour of the dye is not very strong.

Bleaching

Bleaching is much less common than dyeing. It can be used to remove yellow or brown iron staining from jadeite in particular. It is usually followed by dyeing. The processes can be uncertain and perhaps even dangerous but it should at least be considered when any colourless translucent to opaque ornamental material is being tested, since the purpose of bleaching is to make a material as near white or colourless as possible. It is not easy in every case to determine whether chemical bleaching has been done but at the risk of offending some sections of the trade I would suggest that in most cases these are not the most attractive or expensive of materials.

Coating

Coating was certainly employed for the surface of yellowish diamonds which did not reach the strong fancy yellow but were more off-white. Keeping these Cape diamonds in a stone paper with a light blue liner minimizes the yellow colour and a coating of pale blue on the back facets will have a similar effect. Over the centuries the blue has been supplied by foiling or by the application of a copying ink (blue) pencil to the back facets.

The disadvantage of surface coating is that it can easily be damaged and is not too difficult to detect once its existence is suspected. Wax coating of turquoise or lapis is generally intended to improve the colour by deepening it, which is why some dealers in lapis keep their rough material in water. To some extent coating can protect the surface.

Detection is easy with a 10x lens or microscope: small bubbles can sometimes be seen in the substance used for the coating. Damage to the coating can also easily be seen.

Some ornamental materials such as gold on quartz crystals used to give the interesting and quite attractive 'aqua aura', show iridescent colours. Moreover, the quartz crystal forms are not those of beryl which, with blue topaz, is the material imitated.

Impregnation

Impregnation with wax is done with lapis lazuli and turquoise to improve colour and lustre, strengthen the specimen or even enhance transparency. Wax impregnation can be detected by the use of a thermal reaction tester (hotpoint) and a lens or microscope to show concentration of the wax or other material in cracks or in the interstices of carvings. Raman spectroscopy will also indicate the presence of polymer as well as wax surface treatment.

Heating

The heating of some well-known gem species in order to improve their colour – as in the case of aquamarine whose colour is brightened (though rendered more metallic, to my mind) – has been going on for so long that it is taken for granted by both the trade and customers. Aquamarine when mined is often a greenish blue and the heating process may produce a more obvious blue colour. In this case the colour produced by heating is stable and the stone will neither discolour or fade.

Blue zircons seen in jewellery are not this colour to begin with but occur as reddish-brown crystals which may be heated to give not only blue but colourless and golden yellow finished stones. The golden and yellow zircons are stable and unaffected by any heating which may take place later on, perhaps during the setting process, but the colourless and blue ones may develop a dark stain which will spread over the stone if the heating is prolonged. Some will, in fact, discolour over time in general wear but I have not encountered any examples of this for some years now.

Probably the most notable effect of heating on a corundum variety is the fine blue produced by heating Geuda sapphires from Sri Lanka. The material before heating is translucent white. It is not possible to tell whether or not a blue Sri Lanka sapphire has been heated from Geuda material. The best account of this is given by Themelis in the *The Heat Treatment of Ruby and Sapphire*. He describes how in the late 1960s natural blue sapphires with a chalky greenish to bluish fluorescence were appearing at gemmological laboratories; their appearance and behaviour suggested that they could be synthetic. Most if not all this material had been heat-treated in Thailand and the gem trade was seriously affected as far as blue sapphires were concerned since untreated Geuda material was worthless, some having been used as garden path borders and similar low-value items.

Themelis explains that the name Geuda is used for white or colourless, milky, semi-translucent or semi-transparent corundum with included rutile (contributing the titanium necessary, with iron, to provide the blue colour). The name 'diesel' is often used to denote an oily appearance. One example Themelis gives describes the heating of this material for 60 minutes at 1,800°C. The resulting colour was blue, ranging from light to a most beautiful cornflower blue. Some specimens produced no colour while others developed patchy blue coloration and intense colour zoning. Some Geudas developed a slight lilac/purplish overcast coloration due to traces of chromium.

Heating of very dark blue sapphires has been done successfully, but sapphire is not very rare and it may not be economic to treat any but the more promising material.

In general it has been reported that temperatures of less than 1,000°C have been used for the heating of some gem materials. This treatment is low-temperature and unsuitable for some minerals. Corundum may be colour-enhanced by low- or high-temperature treatment and diamonds may have their colour altered by a combination of high temperatures and high pressures (HP/HT treatment).

Heating may often improve colour by lessening the effect of unsightly inclusions, such as excessive rutile needles in some rubies or sapphires. Sometimes the corundum gems may be heated to seal fractures, sometimes with the addition of chemicals. Heating can sometimes produce glassy materials which can fill surface-reaching cavities.

Detection of heating needs considerable familiarity with the inclusions found in the major gemstones, since they are often altered by the process. Any sign of a 'halo' surrounding a solid inclusion usually shows that it has been 'exploded' by the heating but the gemmologist has to learn the forms of the major solid inclusions to recognize the changes they may have undergone. Furthermore the alteration of solid inclusions may have taken place during the growth of the crystal and not as a result of the activities of man.

As far as heat-treated blue sapphires are concerned the following signs suggest that a stone is not in its natural state:

- glassy areas (where 'fingerprints' of growth liquid formerly existed)
- very fine, dot or breadcrumb-like inclusions of rutile
- pock-marked facets and edges
- pock-marked girdles
- chalky blue-green fluorescence with or without an absorption band at 450 nm

As well as altering solid inclusions, the heating process may also give rise to colour concentrations (some blue sapphires develop a spotty colour when heated). Sometimes the response of a specimen to ultra-violet irradiation may alter after heating, but this is not always a conclusive test.

In the case of some gem minerals heating can only be detected by the use of more sophisticated techniques than those available to gemmologists. Raman spectrometry, among other tests, may show when, for example, a tanzanite or tourmaline has been heated.

However, the trade and gemmologists have long taken it for granted that some gem species like these are heated. They do not usually have solid inclusions whose appearance may be altered by the heating, so in general there is no need to pursue whether or not a stone has been heated. In some specimens of diamond or corundum it is very hard to detect treatment. The same can be said of some of the transparent varieties of quartz (amethyst, citrine and smoky quartz, for example).

A good general rule is that the more important the stone, the more it will need a certificate saying whether or not (in the opinion of a reputable gem-testing laboratory) it has been heat treated. Readers who are familiar with the jewellery sales catalogues of the major auction houses will have noticed over recent years that many rubies and blue sapphires carry a note with their catalogue entry saying something like 'no evidence of thermal treatment'. Such entries are most likely to be found in the last pages, since this is where the more important items are described.

Diffusion

While it has perhaps not caught on extensively, the practice of colour enhancement by diffusion needs to be borne in mind by those dealing with blue sapphires in particular. It can usually be recognized by the comparatively shallow penetration of the colour into the originally pale stone. The GIA summary points out the difference between internal diffusion, in which movement of pre-existing atoms takes place when the treatment is applied (the way in which thermal enhancement works) and diffusion treatment, in which atoms giving colour are introduced from an outside source from which they enter and pass through the specimen.

Surface diffusion refers to movement of atoms only at the surface of the specimen, but is now considered incorrect. The term 'lattice diffusion' has been used to describe the movement of colour-causing atoms into the body of the specimen, whether it goes through the whole of the stone or whether it is only shallow. The details of the processes are more fully described in a paper by Emmett and Douthit, *Understanding the New Treated Pink-orange Sapphires*, available from Pala International on http:palagems.com/treated_sapphire_emmett_htm. A summary of ongoing experiments by the same two authors is *Berryllium Diffusion Coloration of Sapphire: A Summary of Ongoing Experiments*, available from the American Gem Trade Association, on http://agta.org/consumer/gtclab/treatedsapps04.htm.

A number of elements have been used in the diffusion process, including the light element beryllium, titanium and chromium.

Most stones treated by diffusion look odd in ways that the lens and certainly the microscope will confirm. While many examples of surface-treated blue sapphire crystals feel rough and look 'sugary' under magnification, the faceted stones obtained from the crystals need more detailed examination. With a hand lens the presence of an abnormally large number of fractures on the facet edges can easily be seen. Also with a hand lens, the colour will be seen to extend only a short distance into the specimen. The techniques of EDXRF, scanning electron microscopy and energy-dispersive spectrometry (SEM-EDS) can detect unusual concentrations of the diffused elements.

When light elements are used for diffusion the colour produced is more likely to extend right through the stone and their presence may not be shown by standard chemical analysis. When the colour does not permeate the entire stone the technique of immersion can be combined with magnification. Diffusion of some of the light elements into corundum needs very high temperatures and these are likely to cause more damage to facet edges and surfaces than heavier elements. A recently recognized technique is laser ablation-inductively coupled plasma-mass spectrometry (LA-ICP-MS) or secondary ion mass spectrometry (SIMS), which is further described by McClure et al. in a paper in *Gems & Gemology*, 38(1), 2002.

Irradiation

Irradiation can be used to alter the colour of a number of gem species. In general it is a less satisfactory form of treatment than heating since in some cases the change may be reversed by quite low energies, even those present in daylight. Furthermore, the techniques may be more complicated and need specialist equipment – health and safety considerations also need to be taken into account.

Electrons, neutrons or gamma rays can be used to alter the colour of the gemstone. The process commonly involves the creation of colour centres (a process which can be reversed) and the colour produced may be shallow or extend throughout the specimen. Heat as well as light may reverse the process. In the case of some important gem species the irradiation process is combined with subsequent heating. This is particularly true of diamond.

Amethyst may lose some though not all of its colour on prolonged exposure to bright light; some yellow sapphires lose most of their colour under similar conditions. It is not possible to tell in advance whether or not a particular yellow sapphire will fade.

On the other hand some treated stones retain their colour. Pink topaz has a stable colour that may have been gained from an original brown material, either after mining or during geological time. Some heat treatment is taken for granted: pale aquamarine is heated to deepen the blue and the dark blue topaz now familiar on the markets is irradiated. In both cases the colour is stable.

Treatment may or may not need to be disclosed (if the seller is aware of it). In the case of aquamarine and blue topaz treatment is taken for granted and disclosure is not necessary. Nor is it necessary for the different colours of zircon (yellow, blue, colourless).

In the early days of treatment a stone could become dangerous to wear or handle if subjected to some types of radiation. Radium-treated diamonds are mentioned in every textbook but there cannot have been very many examples – those that exist are still radioactive and can be identified by a Geiger counter – the diamonds are a rather subdued green. It is possible that such a stone could turn up during valuations. Radioactive stones are very unlikely to be found on a large scale, although from time to time there are newspaper scares – topaz seems to have been the most affected by rumours. Speaking generally, topaz irradiated by gamma rays should not become radioactive, though specimens treated by neutrons in an atomic pile may do. Impurities from the polishing process are more likely to be the source of radioactivity rather than the stone itself.

The earlier edition of this book (1983) described some examples of what turned out to be one-off treatments since either the species treated were either too uncommon to catch the public imagination (whatever the colour might be) or the treatment was too expensive or unpredictable. Quartz, for example, is irradiated to give a smoky colour but naturally coloured smoky quartz is easy to obtain.

The detection of irradiation is not always possible except in cases where the colour gives rise to suspicion. Yellow sapphires are some of the most difficult stones to test. The very fine dark blue Maxixe beryls, first discovered in their natural state in the early twentieth century and later imitated by Maxixe-type beryl in the 1960s, is perhaps the most famous example of the snares of irradiation. This particular species allows the gemmologist to detect the treatment using conventional techniques but in general irradiated stones need such techniques as UV-Visible-NIR to detect irradiation treatment; Raman spectroscopy can also be used.

The eye and experience should not be disregarded either in the case of irradiated stones or specimens which have been colour-enhanced in any other way. In the case of diamond, irradiation produces colours rarely if ever seen in nature (diamond, above all, is the gem species in which almost every colour is seen at one time or another).

Sputtering

A paper in *Gemmologie*, 52(1), 2003, summarizes the present state of the sputtering of gemstone surfaces with thin metallic films. Such treatment (which cannot be very common) is recorded for quartz and topaz in particular but also glass and synthetic spinel. Sputtering may not merely alter or improve colour but also in the appropriate circumstances produce phenomena including adularescence (the moonstone effect).

The detection of sputtering (which has to be borne in mind at an early stage when an unusual specimen is encountered) is best with a microscope. The presence of a distinctly metallic lustre should put investigators on their guard and unexpected body colours may also be notable.

The paper illustrates some examples of sputtering, including a pale yellow and a pink topaz and the same species coloured a bright orange. Structures in the thin film sputtered on to a glass include fine fissures and circular patterns.

CLARITY ENHANCEMENT

The idea behind clarity enhancement is to diminish the effect of inclusions so that the stone appears more glassy. In the opinion of some, myself included, colour is in fact more attractive when light is scattered from inclusions but it does, I admit, depend on the stone. The reflective power of the surface of a diamond, for example, is great enough to overcome some at least of the scattering of light from the inclusions sometimes to be found, while in emerald, with a much lower lustre, the effect of inclusions does sometimes prevent the stone from appearing too much like glass.

The treatment usually involves the filling of surface-reaching fractures with glassy or polymer material, sometimes oil or wax (oiling of emerald is so widespread that all stones reaching the market from the mines are taken to be oiled).

Detection is usually possible with a lens or microscope; in emerald and diamond the flash of an alien colour is highly indicative of filling (this is seen best when the filled area is viewed in a direction parallel to it). Gas bubbles in the filling material are quite easily identified, as are flow structures.

Clarity enhancement is also well known in amber: specimens are often formed from small fragments which are hydraulically compressed to give one specimen. Unsightly inclusions in diamond can be removed with a laser, although multiple tracks (by no means unknown) do give the treatment away.

Jadeite can have iron stains removed by chemical treatment (bleaching), as we have already seen; this is a form of clarity enhancement.

CHAPTER THREE

Gem-Testing Instruments

Gem testing is an increasingly high-level set of techniques and interpretations. A brief review of some of the instruments used in gem testing will help to clarify parts of the descriptive text.

While Raman spectroscopy and Fourier transform infra-red spectroscopy are frequently used in gem testing, they are not everyday gemmological tests as they have come to be known. While gemmologists can still progress quite far with the 10x lens the instruments of the future are to be found only in national museums and the major universities.

The most important resource available to anyone involved in the examination of gemstones is a memory of images seen under the microscope (and from the increasingly expensive literature). The Internet has helped a great deal here and its potential seems limitless.

The 10x lens is the standard tool for the grading of polished diamonds by eye and a description of such a specimen is always taken to be based on eye and lens observation for clarity (details of grading are not covered by this book). It can give a useful first impression but the relatively high magnification means that images move very quickly out of focus (this is a good reason for not using a 20x lens). The outside of a stone is more easily examined under low magnification than the interior; items such as extra facets and damage to girdles are not hard to find.

However, if a microscope is available, it is by any standards the best gem-testing instrument. While a petrological microscope (designed for the examination of rocks in thin section) can be adapted with polars and a rotating graduated stage to give an apparently all-purpose gem-testing instrument it is usually more convenient to use a simpler model and ensure that the working distance is adequate. The majority of microscopes on the market are made for biological examinations and for this reason the working distance is usually far too small to accommodate a relatively large specimen such as a polished gemstone, let alone a crystal.

The next question is whether or not a stereoscopic binocular instrument is chosen in preference to a monocular one. A further choice may well be between a horizontal microscope and a vertical one. There is no doubt that the stereoscopic binocular microscope is the best instrument for examining inclusions. The built-in lighting allows the specimen to be examined from the side (dark-field illumination) as well as by the more customary transmitted light. Most microscopes of this kind are also fitted with an additional light source to provide daylight-type illumination for the examination of diamonds. The performance of either type will be enhanced by the use of additional lighting which can be moved into strategic places – fibre-optic sources are particularly useful in this respect.

The advantage of using a microscope with a horizontal body tube (the eyepiece is angled towards the observer) is that a transparent cell can be used for examination while the specimen is immersed. This cuts out most of the reflections from the specimen's surface, which in other circumstances hinder the observation of what is inside. One small problem can be disturbance of the immersion fluid when someone moves the working surface.

A monocular microscope of the old kind allows the ocular (eyepiece) to be removed to accommodate a hand spectroscope (see below).

Long- and short-wave ultra-violet radiation (LWUV and SWUV) is very useful for the identification of some synthetic materials. Particularly notable is the response of colourless synthetic spinel to SWUV, producing a strong sky-blue glow which is not seen in glass, nor in natural diamond – although some synthetic diamonds do respond in a similar way to SWUV. Chromium-rich materials usually respond well to LWUV – when they respond particularly well they may rightly arouse suspicion that the amount of chromium in the specimen has been artificially controlled.

The use of a refractometer is well described in a number of gemmology textbooks; here it is particularly useful to note that many artificial gemstones will give no readings on the standard refractometer, owing to limitations imposed by the refractive index of the glass and/or the contact liquid used. Over the years the range of possible contact liquids has contracted considerably as a result of health and safety concerns. None the less this is still a major gem-testing instrument; if a gemstone is a true synthetic it will have the same refractive index and birefringence of its natural counterpart, with the exception of flame-fusion-grown spinel.

The dichroscope's workings are best examined in a gemmology textbook: one example of its usefulness in the identification of artificial gemstones can be when it identifies the position of the optic axis in flame-fusion-grown ruby.

Testing for specific gravity (SG) is also explained in gemmology textbooks. Some artificial gemstones, mostly the very rare ones, have a notably high SG and their heft (approximate weight in the hand) rules them out as serious threats in testing.

References to spectroscopy in most scientific texts mean ultra-violet, infra-red or X-ray fluorescence spectroscopy. Gemmologists are able to use what used to be called a direct-vision spectroscope; the name means that observations are made by visible light. This is a particularly useful and easy instrument for the detection of those artificial products whose colour is provided by doping with (the addition of) rare earth elements (REE), many of which show a fine-line absorption spectrum in visible light.

Admittedly using a spectroscope needs practice and, above all, appropriate lighting. The eyes of the observer must be dark-adapted (it is amazing how often this precaution is neglected) and this means submitting to a darkened room for at least twenty minutes before observations are attempted.

Specimens doped with chromium (ruby, red spinel, emerald, alexandrite and green jadeite) will all show elements of the chromium absorption spectrum. Especially well-defined absorption or emission (coloured) elements of the spectrum suggest an artificial product. Many diagnostic spectra need practice to see: one example is the 415.5 nm band in the yellowish Cape diamonds which are usually chosen for colour enhancement.

The Chelsea colour filter was developed to assist in the recognition of synthetic emerald. It allows the transmission of red and green light only. The green of emerald is caused by a small amount of chromium, and crystal growers can of course control the amount used; in nature iron is also usually present. When viewed under a strong light natural emeralds will transmit a strong red through the filter, unless an appreciable quantity of iron is present. Iron can be completely excluded from the starting materials of synthetic emeralds and the stones in consequence appear a very bright red. Even today this can be a useful guide but gemmologists cannot depend on the filter alone; other tests are necessary.

Most gemmological associations sell and give advice on testing instruments.

CHAPTER FOUR

Photography

I n general gemstones can be tested by combining a few reasonably simple instruments: a microscope, a refractometer, a spectroscope, a dichroscope, a polariscope, colour filters and UV sources. All these are readily available but the expertise needed to get the best from them depends upon acquiring a mental library of effects which can be referred to when needed.

While such tests can usually distinguish between one gem species and another, they are not usually sufficient for establishing the true identity of diamond and diamond-like materials; these need more sophisticated instruments which only gem-testing laboratories can provide (and then not all of them).

Treated and synthetic products may also need additional instrumentation and skill. Infra-red spectroscopy is needed to ascertain the nature of some polymeric materials used to fill fractures. Though specific gravity is not one of the tests usually needed for making these kinds of distinction, readers should remember that the chemicals used in one type of SG determination are not longer available without full laboratory safety precautions being observed.

Details of the operation of gem-testing instruments can be found in Peter Read's *Gemmology*.

Familiarity with inclusions is the most vital resource the gemmologist can have. A chapter on inclusion photography formed part of the previous edition of this book, and while the information given there is still valid, the coming of the digital camera means that your photograph can now be sent anywhere in the world as quickly as you view the image yourself.

A microscope which will accommodate a 35 mm or other camera with a removable lens (supplied with the microscope) can get a very good picture using magnifications up to 100x (a general figure – often 20–30x is enough). A zoom lens can be useful and a horizontal microscope allows the specimen to be immersed for the purpose of limiting distracting reflections from the specimen's surface.

Different lighting conditions should be available from your microscope: dark-field illumination where the stone is lit from the side rather than by transmitted light from below is often the best way to reproduce inclusions and it is very useful to have a mobile fibre-optic light source at hand.

Vibration can cause problems, and while an optical bench may not be necessary you might very well want to discourage a colleague from standing by your side with a hand on your table.

Some inclusions are much easier to reproduce than others. The well-shaped and bold-edged gas bubbles characteristic of glass and flame-fusion synthetics are comparatively easy to show, while flat fingerprints (liquid inclusions especially typical of corundum) are much harder.

Many cameras and microscopes on the market can fit together satisfactorily; if not there are several types of adaptor which should be appropriate for almost any eventuality.

Some specimens with large inclusions may cause problems. With a camera the depth of focus can be increased by using a shorter-focus lens, a more distant subject or a smaller aperture. The microscope cannot provide these conditions but a thick inclusion can be photographed at a lower magnification, all of it in focus, and the resulting photograph can then be enlarged. This can also be done with a transparency. However, enlargement will not increase the detail in a photograph but only enlarge it.

Changing the aperture to improve definition is easy with the traditional type of camera in which the iris can be closed down. With a separate camera body the photographer will have to use a separate iris below the objective (they can easily be found in camera shops). Experiments with a small hole in black matt card held in front of the objective may be useful.

The choice of the most appropriate film speed (expressed in ASA or DIN number) is clearly critical. When the level of lighting is low a relatively fast film, consistent with good quality, should be used, perhaps 200 or 400 ASA. Lighting conditions can vary a great deal with the angles chosen and the specimens themselves will reflect light from facets and be uncooperative in many other ways. Their diaphaneity (transparent, translucent, opaque) will also vary and their different colours will need individual attention.

A medium or fast film will cope with most conditions, although a fine-grain ASA film will often be the best for transparent specimens. Processing as soon as possible after exposure will ensure the best results.

Some colour films may show a colour bias and this has to be taken into consideration. If your photographs are to be reproduced in a book this also needs to be checked as printers may also have some idiosyncrasies. It is worth remembering that colour bias may not seriously affect the reproduction of inclusions, as this is not advertising work.

It may be thought an insignificant point but holding the specimen during examination can pose problems, as stones tend to spring away from the blades. Some microscope tongs tend to come adrift from their moorings when set up for examination.

Since the previous edition of this book, digital cameras have made the creation of images so much easier – although one does have to be careful about the quality, which can be disappointing.

CHAPTER FIVE

Diamond

Before you begin this chapter, abandon all preconceived ideas about synthetic diamonds. They are with us and have been for some time. Looking at the diamond picture today we are faced with high-pressure/high-temperature grown stones and now with the more easily grown CVD diamonds. Both will be discussed in this chapter. We shall follow a sequence which we shall use when considering other gem species. Beginning with a general survey we shall then examine the properties of the species and testing in increasingly greater detail as the chapter proceeds.

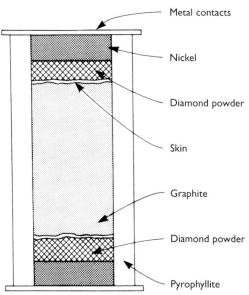

Metal contacts

Nickel

Diamond powder

Skin

Graphite

Diamond powder

Pyrophyllite

Fig. 2 The pyrophyllite container used in the belt system for growing synthetic diamond grit

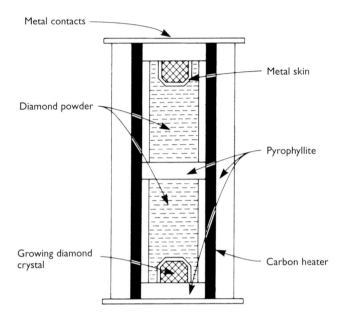

Metal contacts

Metal skin

Diamond powder

Pyrophyllite

Growing diamond crystal

Carbon heater

Fig. 3 The pyrophyllite container used in the belt system for growing synthetic diamond crystals

In one sense it is impossible to tell the public anything about diamond that it does not already know. Just as there are no bad drivers, so everyone can spot a 'wrong' stone masquerading as diamond. While everyone knows about diamond's unique hardness and how it can be polished to give combinations of brilliance (reflection) and fire (dispersion of white light into its component spectrum colours), fewer realize that in the geological context diamond is less rare as a gem mineral than emerald, topaz, garnet and, above all, ruby. Diamonds are in fact relatively plentiful and this is why, since it costs just as much to mine a diamond which will never be used ornamentally as another that will make a fine jewellery stone, prices have to be maintained.

Because of this several complete grading and pricing systems have become established, and while it is not within the scope of this book to discuss the technique of diamond grading it will become clear that the synthesis and treatment of diamond is closely related to the way in which the manufactured or colour-enhanced stones are offered for sale.

Before considering those properties of diamond which clearly distinguish it from other materials, we have to note some aspects of its chemical composition. Although diamond is carbon (and the only gemstone to be formed as a chemical element rather than a compound) it may contain traces of other elements. The most significant is nitrogen, which when present may occur in one of two different ways, and another major one is boron. The presence or absence of nitrogen or boron and the form nitrogen takes are the properties upon which diamond classification is based.

THE CHEMICAL AND PHYSICAL NATURE OF DIAMOND

The Diamond Types

Put simply, Type I diamonds contain nitrogen and Type II diamonds generally do not, although some examples may show very low concentrations.

TYPE I DIAMONDS

In Type Ia diamonds nitrogen is found either as pairs of atoms (Type IaA) or as larger clusters with an even number of atoms (Type IaB). A very high proportion of large, clear natural diamonds – estimated at 98 per cent – belong to Type Ia. They may range from nearly colourless to yellow, but not a deep canary yellow; some may be grey or brown.

Type Ib diamonds are very rare (perhaps 0.1 per cent of all known examples). In these stones the nitrogen atoms are isolated and the colour is usually a deep yellow.

Type I diamonds commonly show a blue fluorescence under long-wave ultra-violet radiation (LWUV). Examples include the yellow so-called Cape diamonds, which also give a diagnostic absorption band at 415.5 nm which can be seen with a hand spectroscope. This band is of great significance to gemmologists since its persistence in a bright yellow diamond shows that the stone has almost certainly been treated to improve its original colour. We should bear in mind that treatment of any kind, in any material, is aimed at improving on nature. Students of the very extensive literature on diamond will soon find references to the N3 absorption band and it is this band that is meant.

Type I diamonds also absorb in the infra-red (IR) and ultra-violet (UV) regions. These tests require equipment which is not usually available outside laboratories, but on the other hand it may be possible to see with a hand spectroscope an absorption band at 504 nm in Type I diamonds inclining more to brown than to yellow. Some of these stones may fluoresce a greenish colour. No Type I diamond is electro-conductive.

TYPE II DIAMONDS

Type II diamonds contain very little, if any, nitrogen and are good conductors of heat (which can be significant in tests which compare the thermal conductivities of unknown stones). Type IIa diamonds are colourless and transmit in the UV region below about 225 nm. Type IIb stones are most often blue, owing to the presence of boron. Type IIb examples are rare (less than 0.2 per cent of diamonds): they are good conductors of electricity and when irradiated by SWUV they may show a blue phosphorescence. They are able to transmit down to about 225 nm. Hand spectroscopes cannot be used in testing since there is no absorption in the visible region beyond an occasional example in the red, which is of little use in testing.

The identification of diamond types is important in the detection of treated stones, as we shall see later.

Physical and Optical Properties

Diamond's unique hardness is based on an atomic structure which also allows a perfect and relatively easy cubic cleavage. This term is used by mineralogists, diamond polishers and lapidaries (cleavage is not peculiar to diamond) to describe a way of breaking which leaves relatively smooth surfaces behind. In diamond cleavage may be started by a sharp knock so that care has to be taken with diamond-set jewellery. For the gemmologist internal signs of cleavage help in identification – they appear as characteristic rainbow-like colour patches.

While these tell-tale signs of cleavage can be found in other gemstones, their combination with diamond's other properties help in testing. The atomic structure which contributes the hardness and cleavage allows the diamond polisher to prepare a very brilliant, near-optically flat surface on the finished stone. This cannot be achieved so successfully on diamond's major imitators. However, surface assessment by eye alone cannot stand up as a serious test, however great the experience of the tester.

Reflectivity can be measured by a simple instrument making use of an infra-red source of 930 nm. Results are usually presented in the form 'diamond or not diamond', without stating what a non-diamond specimen actually is, although one early and quite successful instrument did in fact diagnose quite a wide range of other species from their surface reflectivity.

The SG of diamond is usually very close to 3.52 and the refractive index is 2.42. The dispersion which is so prominent a feature of brilliant-cut diamonds is 0.044, quite a low figure for a material with so high a refractive index.

Diamond's fluorescence under different sources of energy (ultraviolet rays or X-rays) is one of its most interesting characteristics and we shall note some examples later. At this point, however, it is worth remembering that many diamonds fluoresce a fairly characteristic sky-blue under LWUV. This is often taken to be diagnostic for diamond but the truth is less simple. It is true that a piece of jewellery set with many, usually small, diamonds, when examined under a UV source will show an inconsistent fluorescent picture with some stones responding with a blue glow and others remaining inert. Any piece in which all the small stones fluoresce with an even glow should be regarded with great suspicion as diamonds respond very inconsistently to this type of energy.

Specimens which remain glowing after the irradiating source has been turned off are said to phosphoresce but this is not in general a feature of diamond.

Diamond also has the capacity, again owing to its compact crystal structure, to be an excellent conductor of heat. As with its reflective properties, a simple battery-powered testing device has been devised. A copper probe is placed on the surface and registers 'diamond/not diamond'. As some results can be ambiguous (which is also true of the battery-powdered reflectivity meter), several attempts should be made. The great advantage of the thermal conductivity tester is the small size of the probe, which can be placed on very small specimens in complicated pieces of jewellery set with many questionable 'diamonds'.

The ubiquity of diamond and the need to grade significant polished stones means that the diamond trade is not always geared up to testing every doubtful specimen. Because of this, gem laboratories see only a small proportion of specimens which might cause trouble somewhere along the line – perhaps years after their first appearance in the trade.

A microscope is the only gemmological instrument which can generally give an authoritative verdict on most diamonds, but skill in its use takes time to acquire and diamond testers need to develop mental pictures of characteristic inclusion scenes. Diamond, unlike many other gemstones, shows no liquid inclusions and those solid ones which do appear (they are not uncommon) are harder to diagnose than photographs in textbooks sometimes suggest.

Now that synthetic gem diamonds have been on the market for some years (polished synthetic diamonds have been turning up in laboratories since at least 2000), gemmologists and diamond graders have to learn a new set of microscope images. First, a generalization: many synthetic diamonds show metallic flux inclusions which do not really look like any of the solid inclusions to be found in the natural stones. Those gemmologists who have become familiar with inclusions in flux-grown emerald and ruby will more easily recognize flux in diamond by its silvery appearance under reflected light. We shall look at them further later.

It often requires a mental gear-change to remember that not all diamonds are colourless. Apart from the yellow Cape stones already mentioned, which are not a very strong colour, 'fancy' diamonds can show a range of strong colours, of which yellow is the commonest and pink to near-red the rarest and the most desirable. It is possible today, and has been for many years, to alter the colour of some diamonds (in particular Cape stones) to make them much more presentable. The colour most often achieved is a fine bright yellow (canary yellow) and any fine example turning up today needs to be tested unless it is accompanied by a certificate from a recognized gemmological laboratory. Blues are not deep and often appear metallic while greens are dark and nothing like the green of emerald.

The artificially induced colours of diamond are stable; specimens will not fade or discolour. Their overall appearance gives no clue to their history. Their rarity, however, keeps prices high and any stone unaccompanied by a certificate should be sent for testing.

THE COMMON SIMULANTS OF DIAMOND

Once more commonly known in the trade as paste, glass is the commonest imitation of diamond, as it is of all transparent natural gemstones (and some translucent and opaque ones too). But the aim of those attempting to manufacture a satisfactory simulant of diamond is to reproduce the brilliance and high dispersion of the original, and this cannot easily be achieved with glass. Crystal growers in the twentieth century therefore turned their attention to materials which might at least go some of the way. These materials were not at first grown with this aim in mind but rather for research or industrial purposes, but while some of them appear too garish to be serious diamond simulants, others have caused some alarm in the trade, at least on their first introduction and before the gem laboratories came to the rescue.

Glass

Glass is softer than most natural gemstones (although there are some hard glasses), and cannot show the cleavage of diamond as it is not crystalline. Glass shows a very pronounced shell-like (conchoidal) fracture, so gemmologists look with 10x lens or microscope at girdle and facet edges for fractures. Glass is a poor conductor of heat and diamond a very efficient one so the thermal conductivity tester will easily separate them.

The interior of glass is characteristically swirly, as if the materials of which it is composed had not adequately mixed – which is close to the truth. In addition and more easily distinguished are the well-rounded and bold-edged gas bubbles randomly distributed through the specimen. Isolated bubbles of this kind are not seen in any natural stone, where any gas bubbles form one phase in a multiphase inclusion of solid, liquid and gas.

Glass can be manufactured to show any colour, so this is not a reliable guide to the identity of a particular specimen. Glass can also be faceted to show quite a high dispersion, which can be enhanced by the addition of lead. Here again the eye can be deceived.

When specifically gemmological tests are employed it is easier to establish that a specimen is not a diamond; if diamond is expected it does not matter to the tester what the substitute actually is. When a refractometer is used, diamond cannot give a reading while glass may or may not, depending upon its composition and the way in which it was fashioned. Especially when moulded, it will probably give a table which is insufficiently flat for the refractometer to give a reading. When a reading can be obtained it will often be between 1.50 and 1.70. As glass is not crystalline there can be no birefringence.

Glass does not show a consistent fluorescence or phosphorescence. Occasionally glass doped with rare-earth elements (REE) may show an unexpected absorption spectrum but this is not very common and in any case the comparative complexity of many REE spectra is unlike anything ever reported from diamond. Just as the possible refractive index covers a wide range so does the SG, so testing for this property is not always helpful; it can also be slow and prone to error.

Synthetic Rutile

One of the early examples of grown diamond simulants is the highly dispersive and rather soft titanium dioxide, rutile, which is common in nature. When faceted, while the dispersion is quite remarkable and far exceeds that of diamond, the quite obvious birefringence, shown by the doubling of opposite facets and inclusions, at once rules diamond out. The colour of artificial rutile, although meant to imitate colourless diamond, succeeds only in achieving an off-white with a hint of yellow. Readers will immediately think, 'But this is quite like the appearance of some Cape diamonds.' This may be partly true in some examples, but growers would not normally be content with producing stones which look like the lower qualities of diamond. Rutile does not show the Cape absorption at 415.5 nm.

Rutile comes more into its own when doped with elements which give a golden brown or blue colour. While these are unusual, the dispersion is far too high for diamond and the blue at least too deep a body colour.

If for some reason one does not spot rutile's very strong birefringence of 0.287 (as a comparison the birefringences of two natural gem species, peridot and tourmaline, are about 0.018 and 0.036 respectively; diamond shows none, of course), there are always other tests. Before testing it is well to remember that rutile (like diamond) cannot give a reading on a gemmological refractometer nor, unless the specimen is unmounted, can an SG test be carried out. Even then the test is slow and can be prone to error. The reflectivity meter will tell you that rutile is not diamond, but nothing else.

So we have to fall back on the microscope as the first line of investigation. If the birefringence is missed there is not much more to say. We should always remember, though, that small stones give the most trouble and while rutile melée would not often be encountered, examples do exist.

For the record the hardness of rutile is 6.5, the refractive index (RI) 2.62–2.90 and the birefringence 0.287. The dispersion is close to 0.3. There is no fluorescence.

The trade name most in use was Titania, which by convention also denoted the oxide of titanium, but the material was somewhat short lived and not many different names ever accumulated. Rutile was manufactured by the Verneuil flame-fusion process and boules were cheap and quickly grown.

Strontium Titanate

The material is also grown by the Verneuil method but in this case, unlike rutile, a completely colourless crystal is achieved. Nor does strontium titanate show birefringence. It may thus be thought to be more suitable as a diamond simulant. It is now known to have a natural counterpart, the mineral tausonite, but this has no ornamental importance.

As a diamond simulant strontium titanate (commonest trade name Fabulite) shares rutile's drawback of low hardness. It is possible to mark the surface with the point of a needle and for this reason the material has often been used as the base of a composite (doublet) in which the effect of its high dispersion is lessened by fixing it to a crown of an easily available and cheap material, usually synthetic colourless corundum or spinel. These hard substances also serve to shield its softer pavilion from accidental scratches.

Like rutile, strontium titanate has been doped with different elements to give coloured crystals which have sometimes been faceted. Doped versions are rare and can be very attractive; the absence of the pervasive yellow cast helps to improve their appearance and that of the colourless material. Of the two, strontium titanate looks in general more like diamond than rutile does, with one of the doped versions sometimes resembling light pinkish-orange diamond.

While this rare stone can be quite convincing, far more dangerous is strontium titanate melée. The dispersion shown by a group of small stones in a parcel with a dark background is very like that shown by diamond of similar sizes. These are some of the most deceptive artificial gemstones. The doublets already mentioned are joined above or below, rather than at the girdle and the join can be seen under magnification – if it is expected. I have often said that the gemmologist or jeweller often has to 'think composite'. If we do not, the possibility may escape us. This is even more likely now that there are so many other things to look out for: synthetic gem diamonds, evidence of colour

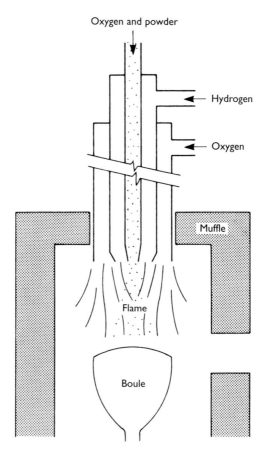

Fig. 4 The tricone modification of the Verneuil flame-fusion furnace. It was used for growing synthetic rutile and strontium titanate

enhancement and new species. Composites do not just fade away: they stay around to make life difficult but interesting.

Like rutile, strontium titanate has given way, as far as the market is concerned, to cubic zirconia and perhaps even to synthetic moissanite.

The properties of strontium titanate can be summarized as: hardness 5.5–6, RI 2.41, singly refractive, SG 5.13, dispersion 0.19, no fluorescence or phosphorescence.

Both rutile and strontium titanate appeared as diamond simulants after the Second World War. Neither really caught on and examples are not plentiful compared with glass. Their softness and dispersion were not in their favour and it is possible that the need to grow more complex crystals for electronic purposes diverted the attention of growers from relatively simple materials for which demand can never have been great.

Synthetic Garnets

Between the advent of rutile and strontium titanate and that of cubic zirconia the synthetic garnets occupied the stage quite effectively. Known most commonly as YAG this group of substances differs from natural minerals of the garnet group in their chemical composition: natural garnets are silicates and the synthetic ones oxides. The link between the synthetic and natural garnets is structural rather than chemical. None of the synthetic garnets has a natural chemical counterpart.

Synthetic garnets are found in both colourless and coloured forms. With rutile and strontium titanate a range of bright colours is obtained by the use of dopants; in general, coloured synthetic garnets are too strongly coloured for use as simulants of fancy-coloured diamonds and the colour range is greater. The colourless form alarmed the trade if the media of the time are to be believed and the trade name Diamonair passed into quite general currency (there were many other names).

The name YAG (there are several analogues including GGG) is the abbreviated form of yttrium aluminium garnet $Y_3Al_5O_{12}$. Colourless stones appear less dispersive than diamond and certainly much less than rutile or strontium titanate. Though hard, YAG not surprisingly lacks the brilliance of diamond. Colourless stones show no fluorescence.

YAG has an RI of 1.83 (as with diamond, rutile and strontium titanate, this figure cannot be obtained with a gemmological refractometer) compared with diamond's 2.42. The hardness is about 8.5 compared with 10 and the SG 4.55 (diamond is 3.52). Comparing these figures makes it clear that YAG cannot seriously compete with diamond when handled by experienced members of the diamond trade or gemmologists. None the less it was a greater success than might have been expected and, of course, specimens are still around – old synthetics never die, nor do they fade away.

Those forms of YAG and GGG which have been doped by the addition of other elements are rather more attractive than the colourless forms. Perhaps fortunately, neither the green of emerald nor the red of ruby has been exactly reproduced but the attempts have given colours attractive in themselves. When the dopants used are rare earths, specimens are quite likely to present a fine-line absorption spectrum unlike that shown by any natural mineral, so the gemmologist should reach for the spectroscope when faced by a bright, hard-looking, inclusion-poor specimen. Not all rare earths will show strong absorption spectra in the visible region, however.

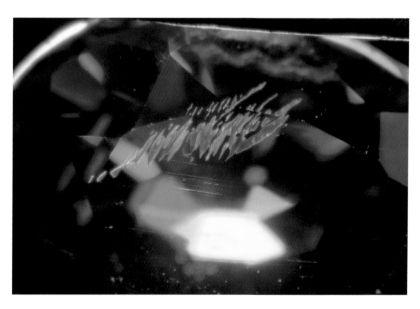

YAG (imitating demantoid garnet). The irregular flux residues bear no resemblance at all to the chrysotile fibres in natural demantoid garnet

YAG (green). Large freeform flux inclusions

A microscope may show inclusions associated with the growth method of the original crystal but it is quite usual for specimens to show nothing under normal gemmological magnification. It may also be useful to hold a specimen over a dark background with a single source of light above and look at the table facet, which should be uppermost. When a cut stone is slowly tilted away from the tester the stone will appear to lose light progressively as the angle of tilt is increased. This does not happen with a properly polished diamond. Both YAG and cubic zirconia differentiated from diamond by this test but previous practice with known specimens is essential. There is rarely any need for the two simulants to be distinguished from one another.

It may also be possible for a round brilliant specimen to be placed with the table facet down upon a small black dot drawn on a sheet of paper. Diamond simulants can be distinguished from diamond by the appearance in them of a ring surrounding the culet (bottom facet or point of the stone).

Properties of synthetic garnets are: hardness 8.5, single RI 1.83, dispersion 0.028 (diamond is 0.044), SG 4.55.

Synthetic garnets may be grown by the flux-melt method but crystal pulling is much quicker, giving virtually inclusion-free crystals.

Synthetic Spinel

Synthetic spinel, a very common simulant of diamond, is easy and cheap to make. It is grown by the flame-fusion technique and fluoresces a bright sky-blue under SWUV. When natural diamonds fluoresce this colour it will be under LWUV; some synthetic diamonds will, however, give the same type of fluorescent response to SWUV. The RI can easily be measured in reasonable circumstances (1.728). The SG will be 3.64. Between crossed polars a striped effect will be apparent.

Synthetic Moissanite

The synthetic form of the silicon carbide moissanite, SiC, has been manufactured for ornamental and gem use; in the 1960s some iridescent though opaque crystal groups were around at gem and mineral shows, but it was only in recent years that a transparent variety was synthesized, in North Carolina, USA. Most of the properties of diamond are quite well imitated and the usual anxiety was reported to be pervading the trade (such reports are usually exaggerated, and in any case, there *are* laboratories).

From the gemmologist's point of view, synthetic moissanite belongs to the hexagonal crystal system and is thus birefringent. Under magnification – the 10x lens serves very well – opposite facet edges appear doubled when viewed in the appropriate direction. The effect cannot usually be seen through the table facet, which would have been easier to detect, but at right angles to it. The setting usually obscures observation across the girdle so care needs to be taken. In many specimens near-parallel needles and stringers may be seen at right angles to the table. Some specimens show rounded facet edges (those of diamond are exceptionally sharp), which are not in themselves a vital clue. There are also some unidirectional polishing lines.

Its gemmological properties are: hardness 9.25, RI 2.648–2.691, birefringence of 0.043, uniaxial positive, and dispersion 0.104, which is more than twice as great as diamond. The SG is 3.22 (diamond is 3.52). These properties can all be tested with a little effort but with any diamond imitation it is usually worth devising a catch-all detector. Reflectivity meters have usually been used to separate diamond-like transparent stones from their more serious imitators YAG and cubic zirconia, although they can only reach a few stones in a piece of jewellery which contains many small ones in hard-to-reach places. These 'lurking' specimens can more often be reached with a thermal conductivity tester, which will very effectively separate diamond from most of its simulants.

A suggestion going the rounds in recent years was that manufacturers of synthetic moissanite might be able to alter the refractive index of their product so that the reflectivity meters gave a diamond reading. (Reflectivity meters do not measure refractive index as such but RI does influence reflectivity.) It might therefore have been possible to repolish the specimen so that the original RI could be read. Rumours of this kind often circulate through the trade and those involved should ask themselves, 'What might the manufacturer gain from all this trouble?'

A synthetic near-colourless moissanite has been heat-treated, which causes a brownish colour to develop across all the facets. Cleaning and hand-polishing the samples with cerium oxide on leather restored the reflectivity to 98 per cent of the non-treated material. If heating forms part of any testing experiment on a suspected moissanite the gemmologist should remember that surface oxidation could occur and keep the level of heating to a minimum. The colour of the surface might undergo alteration.

A thermal conductivity tester will give a diamond reading for synthetic moissanite in any case, so it might be used more to separate diamond and synthetic moissanite from other diamond simulants; a further test to separate the two could well be magnification.

Diamond will sink and moissanite float in di-iodomethane (SG 3.34).

Some coloured moissanites have been grown – green, yellow and blue – but the colours are not very strong. Green specimens I have seen are not like green diamond. A brown moissanite is reported to have been grown in Russia by chemical vapour deposition: the specimen described was opaque.

The first firm to produce synthetic moissanite and sell it (exclusively at first) was C3 Inc. in North Carolina, USA. The same firm, now called Charles & Colbard, produced a tester, which they sold under the name Tester Model 590. Other instruments have followed, which separate moissanite from diamond by scanning the blue and near-visible UV areas of their absorption spectrum. In moissanite there is an intense region of absorption extending down from about 425 nm to the UV region. Colourless diamond on the other hand transmits well down into the UV. A halogen light source directs a beam onto the table facet which reflects it. If it transmits wavelengths from the blue to the UV region the tester gives a visual indication plus a bleep, indicating that the specimen is diamond. If no response is given the specimen will be moissanite, having absorbed this range of wavelengths.

Another instrument, the Presidium moissanite tester, detects the very small current passed by semiconducting materials. The operator receives a signal indicating 'diamond/not diamond'. Synthetic moissanite is not a semiconductor, however, so that any current detected may be caused by impurities in the material.

In a paper in *Australian Gemmologist* (20, 483–5, 2000), the authors reported that the testing of synthetic moissanite was made easier by the Presidium tester. As synthetic moissanite is a semiconductor the instrument is able to sense a forward leakage of current. Other testers are based on the 'breakdown voltage' (which overcomes resistance to an insulator causing current to flow).

The authors found that all synthetic moissanite specimens were detected as such by the apparatus, indicating them by illuminating a bright red window display and a sound alert. Other species caused a green 'test' lamp to be illuminated. False positive synthetic moissanite readings occurred during the evaluation, particularly when the tip of the probe was in contact with metal (such as the setting) and also when a germanium transistor was touched. A synthetic moissanite response was also given by a black electrically conductive industrial quality diamond.

Cubic Zirconia (CZ)

The most effective simulant of diamond followed the colourless synthetic garnets and, like them, was a member of the cubic crystal system. In nature, zirconia is not cubic and in growing the cubic form additives are needed to ensure that the appropriate symmetry is achieved; either yttrium or calcium are added.

The melting point of zirconia is 2,750°C, so no crucible can be used in its growth. Instead the crystals grow within a melt of their own powder, a block of which is placed within a source of radio-frequency induction. The block melts from within, the process being assisted by the incorporation of some pieces of zirconium metal and copper tubes carrying circulating cooling water which enclose it, giving the rather vague suggestion of a skull.

On completion of growth, columnar crystals of zirconia are retrieved. Doping is possible, although there do not seem to have been serious attempts to imitate coloured diamond.

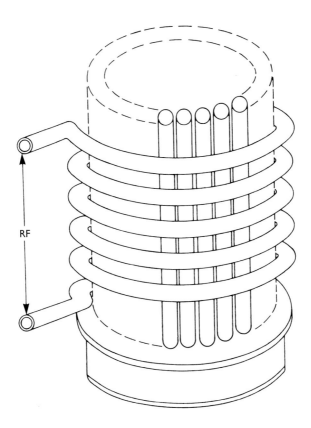

Fig. 5 A skull-melting apparatus set up for use. (RF = Radio Frequency)

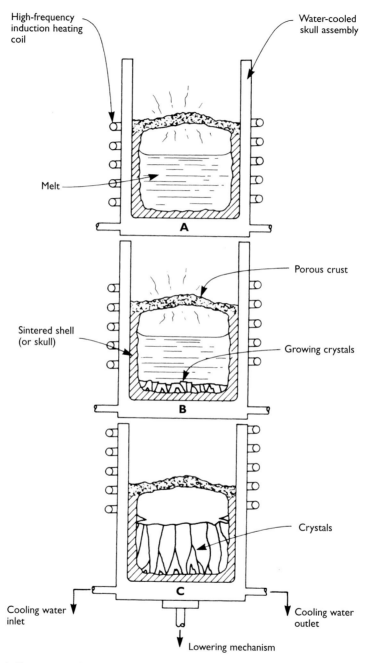

Fig. 6 The process of solidification during skull-melting crystal growth: (A) formation of a porous crust; (B) growth of parallel columns; (C) the whole melt has solidified

The properties of CZ are hardness 8.5, SG 5.95, RI 2.171, dispersion 0.059 (compared to the 0.044 of diamond). These figures are for the yttrium-stabilized form; the calcium-stabilized version has SG 5.65, RI 2.177, dispersion 0.065. In practice, gemmologists do not need to distinguish between the two forms.

Calcium-stabilized CZ may show a distinct yellow fluorescence; the yttrium-stabilized material has been reported to show, on occasion, a greenish-yellow or reddish fluorescence. In both cases the fluorescence shows under LWUV.

Neither type shows any form of inclusion by which it can be recognized, though some yttrium-stabilized crystals have been reported to contain parallel rows of semi-transparent cavities extending into hazy stripes. Calcium-stabilized CZ appears not to show any significant inclusions.

Testing cannot be carried out by the commoner gemmological instruments as the refractive index is too high for the standard refractometer and for the liquid in common use. If the specimen is mounted it will not be possible to ascertain its specific gravity and colourless material will show no distinctive absorption spectrum (CZ doped to

Zirconia made by H. Djevahirdjian is characterized by numerous gas bubbles

give colour will most likely show the fine absorption lines of a rare earth element). CZ often shows rounded facet edges.

The gemmologist is thus forced back upon the reflectivity or the thermal conductivity meter. The absence of any solid inclusions should be a useful clue to CZ but confirmatory tests will be needed.

How long CZ will persist as a diamond simulant is uncertain, though large-scale growth of synthetic diamond is probably some way off. Certainly CZ is not expensive to grow and it can be doped, unlike diamond, to give strong and attractive colours. Analogues such as hafnia (colourless, transparent) have not caught on, owing to the expense of the starting material needed.

SYNTHETIC DIAMOND

In March 2004 a television programme reported the appearance of synthetic diamonds, making the point that the trade and the public will have to adjust to a completely new way of thinking about values and prices. It is too early to forecast the way in which the trade will move, but some very skilled advertising will be needed.

The synthetic and high-pressure/high temperature (HP/HT) diamond problem will affect other gem species. The whole concept of 'synthetic' must not be confused with 'fake' and the long-held belief that any synthetic must be glass also needs to change. My view is that many years will need to pass before synthetic and treated stones find their own niche in the trade instead of always having to lurk behind natural materials.

History

As long as 150 years ago reports of the successful synthesis of diamond had begun to circulate. Some were convincing but with the benefit of hindsight we now know that the products hailed as a breakthrough in the apparently impassable barrier of achieving sufficient heat and pressure together were not in fact diamond – or if they were they were not manufactured by the apparatus used for the experiment. Some were, it now seems, natural diamonds placed there either to assist the progress of the experiment or to give the impression that a dangerous attempt had succeeded and that there was no need to put lives at further risk.

Some of the stories are well told by Nassau in *Gems Made by Man*. There was no difficulty over diamond's composition or structure: both were well understood. The problem was to achieve the growth of diamond rather than another crystalline form of carbon. Interestingly and contrary to what might be thought, diamond is the less stable crystalline form of carbon and graphite the more stable.

The growth of diamond has been the subject of numerous monographs, including a recent study by Amanda Barnard, *The Diamond Formula*. She introduces the history of the many attempts to grow diamond and summarizes those unique properties of the material which have always made growth a problem.

This book is not intended as an in-depth study of all the possible methods by which gem materials can be grown, although it is hoped that readers will be able to gauge how properties limit simple growth methods in the case of many materials and how testing methods have had to be developed to assist gemmologists. For this reason I shall attempt to give a summary of the thinking behind synthesis.

Diamond and graphite are both forms of carbon and both are therefore known as polymorphs of that element. While the best way to create diamond in general is by the conversion of graphite, diamond powder is needed for the creation of gem-quality diamond.

The greatest impetus to diamond growth study in the nineteenth century came from the realization that diamond creation must involve high pressures. One of the first workers in this field, J.B. Hannay, attempted to produce diamond in 1878 by heating various volatile carbonaceous compounds in a sealed iron cylinder of 10 cm diameter and 1.25 cm bore, which was sealed by welding.

Heating for some hours caused the tube to explode at a pressure later estimated at 7,000 atm. This was presumably expected since Hannay estimated that pressure was necessary to transform the carbon present into diamond. Barnard, quoting authorities reviewing Hannay's results, tells us that about eighty such experiments were made but only three cylinders remained after the explosions. Some of the explosions were of a magnitude to threaten Hannay and his workers, but no trace of diamonds was found. After one of the experiments, however, diamond crystal fragments were found, and nine of these are still known.

The crystals available from the apparently successful experiment were examined and details published in *Proceedings of the Royal Society* during 1880. In recent years the specimens have been re-examined and their identity as diamond confirmed. What was not confirmed, however, was their synthetic nature. Nitrogen was found in aggregated concentrations and showed that the specimens were natural diamonds. How they came to be there has been the subject of conjecture ever since, and further speculation is outside the scope of this book.

Similar experiments were conducted by other workers among whom probably the best-known today was Frédéric-Henri Moissan, who gave his name to moissanite. Describing his more elaborate apparatus, Barnard relates its apparent success when crystals of octahedral form were found at the conclusion of one of the runs. These were found to be diamond: they burned in oxygen and scratched minerals known to be hard. The presence of diamond has commonly been thought to be the result of 'salting' the apparatus with crystals already known to be diamond, perhaps by one or more of Moissan's assistants.

Interestingly, Moissan did succeed in synthesizing silicon carbide, moissanite. At that time this substance had not been found in nature, although natural moissanite was eventually reported in 1976 and 1992 (several polytypes exist). Synthetic moissanite cannot easily be mistaken for diamond since it is doubly refractive (anisotropic); the distinction can quite easily be seen when pointed out, and is certainly within the capability of someone with a 10x lens.

Barnard reports the efforts at diamond synthesis made by Sir Charles Parsons, who followed Hannay's method with superheated iron cylinders and sources of carbon. This time the resulting crystals were found to be spinel. There is occasionally debate as to whether or not some of the crystals made by earlier workers were spinel, since both diamond and spinel form characteristic octahedra. This, however, is a matter for the historians of diamond growth.

Reviewing these early attempts one must have a good deal of admiration for the pioneers who undertook such lengthy and sometimes dangerous experiments. In general their ideas were correct – it was only their application which did not succeed at first. Barnard tells us that while they recognized the need for heat in the synthesis of diamond, they thought that it was to assist in the creation of pressure rather than playing an independent part.

Advances in high-pressure and high-temperature physics eventually made possible the successful synthesis of diamond and in this work P.W. Bridgman is a major figure. He devised a press with a container for the materials from which the desired product was to be obtained. The container was made from pyrophyllite, which deforms uniformly at high pressures, thus forming a non-extruding gasket which does not break under extreme conditions. Bridgman's apparatus became known as the 'Bridgman anvil' and the term 'anvil' is often encountered in accounts of diamond synthesis, referring to the pistons which exerted pressure on the container.

Bridgman was able to achieve pressures of 20,000 atm with the use of the non-extruding gasket and, with tapered anvils, up to 50,000 atm. A number of materials were grown by this apparatus, which in time was able to achieve pressures of 400,000 atm.

However, the successful synthesis of diamond from graphite had to wait until after the Second World War when the Swedish company Allmanna Svenska Elektrika Aktiebolaget (ASEA) succeeded in making crystals in 1953. For years there was considerable doubt about who was the first to make synthetic diamond, because in 1955 General Electric (GE) published details of their successful synthesis in 1954. ASEA kept their secret until 1986. According to Barnard they did not realize that the process could be patented as well as the apparatus.

At the GE laboratories in Schenectady, New York, diamond synthesis was the responsibility of Project Superpressure. At this time the term 'belt' came into use, referring to a modification of the Bridgman anvil. The success of the GE team depended on a solution of the pressures and temperatures required. Growth was finally found to have taken place on seeds of natural diamond.

The report states that in December 1954 Herbert Strong seeded graphite with a small natural diamond crystal, covered them with iron foil and placed them in the apparatus which gave pressures of 9,000 atm at a temperature of 1,250°C. The apparatus is reported to have run for one night, following which two synthetic diamond crystals were found, independent of the natural diamond seeds.

A similar experiment carried out on 16 December 1954 by H. Tracy Hall, using the belt system, produced hundreds of minute diamond crystals. The pressure exerted was about 100,000 atm and the temperature 1,600°C for thirty-eight minutes.

GE's report inevitably caused turmoil in the diamond and gemstone world. Interestingly GE, like the Swedes, failed to take out international patents at first, depositing papers only with the United States Patent Office. De Beers, who had established a diamond research project at their Johannesburg laboratories, made a high-pressure apparatus and began experiments in May 1959. Considerable litigation over the use of the belt apparatus ended with De Beers buying limited rights to it via a licensing agreement.

A new apparatus for diamond synthesis came about as a result of Hall's move from GE to Brigham Young University, Utah. Here he developed the tetrahedral press, in which four anvils exerted pressure on the growth chamber, heat coming from an electric current passing through the anvils. Hall's contribution to diamond synthesis research has been considerable, since he worked on the belt apparatus, the tetrahedral press and a six-anvil cubic press.

It is interesting to see some of these devices in unexpected places; some years ago when I was at the National Physical Laboratory of India I saw a tetrahedral press; to give some idea of the size, the top of the machine was about 2m from the ground.

Barnard describes a further apparatus for the growth of diamond, this time from Russia, where workers at the Novosibirsk scientific complex developed a system in which a series of anvils was arranged in a spherical shape with a decreasing number in concentric layers. It is reported that this arrangement achieves even and uniform pressure over the whole of the container in which synthesis takes place. Catalysts include iron, nickel, manganese and their alloys as well as transition metals. Growth has been estimated at more than 5 mg per hour.

GE Diamond

The appearance of synthetic gem-quality diamond on the market was reported by the Gemological Institute of America (GIA) in 1984. The specimens described were three polished stones and five unpolished crystals. The cut stones were polished as brilliants: one was near-colourless, one bright yellow and the third greyish-blue. None of them responded to LWUV but under SWUV they behaved differently, the near-colourless specimen giving a very strong yellow fluorescence and phosphorescence. Neither the yellow polished stone nor the yellow crystal responded to SWUV, but the greyish-blue stone showed a strong and slightly greenish-yellow fluorescence with a similarly coloured and persistent phosphorescence. The blue and near-colourless stones showed a cross-shaped pattern when examined under SWUV.

Such stones were probably experimental and not intended for the trade.

None of the original diamonds showed absorption in the visible region, although, as they were a similar yellow to Cape diamonds but lacked the absorption bands shown by Cape stones at 415.5 nm, visible with the hand spectroscope, gemmologists might have been suspicious.

A common test for Cape stones of importance is to examine the absorption spectrum under cryogenic conditions but the GE stones showed no absorption even then. In one small specimen the band was observed but it was ascribed to the presence of a small natural diamond which had acted as a seed for the host stone.

GE near-colourless and blue synthetic diamonds were electrically conductive, which is a property of Type IIb diamonds. Natural diamonds of Type II are colourless, or blue when boron is present. Aluminium impurities were thought to have been responsible for the electroconductivity in this case. No natural near-colourless diamonds had been found to be electroconductive before 1984.

Metallic iron from the growth process explains the higher specific gravity (around 4 as against diamond's normal 3.52) of the earlier synthetic GE diamonds. The iron was present as inclusions.

Some, but not all, of the blue diamond and the near-colourless one were strongly attracted by a pocket magnet, but the yellow stone was much less affected. It would be possible, if a superconducting magnetometer were brought into use, for GE diamonds to be separated from natural ones on the basis of magnetic properties. In practice an RE magnet provides an adequate means of testing in this context.

When examined between crossed polars no sign of strain could be detected: this is unusual in diamond, which commonly shows irregular light and dark patches during a complete rotation.

These two last properties, with the electroconductivity of the near-colourless specimen and the strong response to SWUV while remaining inert to LWUV, seem to be sufficiently characteristic of GE diamonds to form a diagnosis of them. Certainly a near-colourless diamond without any tint of blue or grey and with the fluorescence/phosphorescence effects described above, will in all probability be synthetic.

In general fancy (strongly coloured) natural yellow diamonds showing no absorption bands with the spectroscope usually fluoresce and phosphoresce under LWUV. Fancy yellows which do not show this response will normally give the absorption band at 415.5 nm.

A yellow diamond showing neither of these features may very well turn out to be synthetic. A near-colourless diamond with no hint of grey, blue or brown and no Cape absorption spectrum is also likely to be synthetic. Any diamond with strong fluorescence under X-rays with strong and persistent yellow phosphorescence also suggests artificial origin since most natural diamonds show a blue fluorescence under X-rays.

Sumitomo Diamond

In 1985 the Japanese firm Sumitomo released specimens of synthetic diamond. These were yellow crystals up to 2 ct in weight and were said by the firm to be for industrial applications, but crystals I have seen are attractive, a rather dark golden yellow and offered as flat, near square sawn shapes. The strong yellow colour is probably caused by the presence of nitrogen. As we have seen nitrogen in the form of disseminated aggregates determines a diamond as Type Ia and these make up a high proportion of natural diamonds. Their colour is yellow or near-colourless with some brown and grey samples.

In nature Type Ib diamonds are very rare so that any yellow Type Ib diamond is very likely to be synthetic. The nitrogen content is less than in Type Ia diamonds and is dispersed rather than forming aggregates. Type Ib diamonds are usually a deep yellow. Type IIa diamonds are usually colourless and their nitrogen content uncertain and hard to predict. Few have been found in nature, although they are less rare than Type Ib.

Type IIb diamonds are also rare in nature and apparently contain more boron than nitrogen. They are blue or grey, sometimes colourless, and conduct electricity.

The synthetic diamond types may be differentiated by infra-red spectroscopy but if you need to distinguish between a Sumitomo, a GE and a natural diamond (which would be a task for a research laboratory) the following points may be useful.

The UV response is different. Type Ia diamonds may fluoresce orange, yellow, green or blue, and they may be inert. Type Ib diamonds may show similar fluorescence colours, whose intensity may vary. GE and Sumitomo diamonds are inert. Under SWUV the same types show a similar but more variable response, while some at least of the GE stones are inert and the Sumitomo diamonds show a moderate to intense yellow or greenish-yellow response.

Neither the GE nor the Sumitomo diamonds show phosphorescence while Type Ia and Ib diamonds either show it in different colours or not at all, and this applies to both LW and SW radiations. Type Ia and Ib diamonds phosphoresce with similar variability under X-rays or show no phosphorescence. GE diamonds show no phosphorescence but Sumitomo crystals show a weak to moderately intense bluish-white phosphorescence.

With a hand spectroscope Type Ia diamonds may show some sharp absorption bands; Type Ib diamonds show none. Neither of the two synthetic diamonds showed any absorption in the visible region.

Although colour zoning can be seen in natural diamonds the Sumitomo stones have shown a deep yellow inner zone and a narrow, near-colourless outer zone. Some variation in the yellow zone has been reported. GIA has reported whitish pinpoint inclusions and opaque black metallic inclusions; the latter arise from the flux used in the growth of the crystal. These inclusions are never present in natural diamond. In general natural and synthetic diamond crystals are polished in such a way that the inclusions are avoided – careful cutting also avoids the veil-like colourless areas observed in some stones – these extend only a short distance inside the stone.

The synthetic Sumitomo diamonds show prominent graining, with one type consisting of sets of lines both inside and outside the crystal and the other occurring only within the stone, consisting of sets of straight lines radiating outwards from the centre of the crystal in four wedge-shaped formations resembling a cross with splayed ends. When the crystal is polished the cross turns to an hourglass shape, which can be seen through the bases of the crystals examined.

Between crossed polars the Sumitomo diamonds show what now seems to be a characteristic 'bow-tie' effect, caused by interference. This effect is always possible in other synthetic diamonds but the Sumitomo products were the first to be reported.

De Beers Diamond

The De Beers Research Laboratory in Johannesburg followed the GE and Sumitomo products with a synthetic diamond of its own. The crystals ranged in colour from light greenish-yellow through yellow to dark brownish-yellow.

The brownish-yellow diamonds were inert to LWUV but gave a moderately strong fluorescence of yellow or greenish-yellow colour with some inert zones under SWUV. The yellow stones were inert under both types of UV. The greenish-yellow diamonds were inert under LWUV and responded with a weak, yellow-zoned glow to SWUV. Phosphorescence was shown only by the greenish-yellow specimens, but when seen it was persistent.

None of the De Beers diamonds showed any absorption in the visible region, but all specimens showed distinct colour zoning (the greenish-yellow stones less strongly than the others) and the hourglass pattern was observed in the brownish-yellow and yellow stones.

In the brownish-yellow stones dense clouds of whitish pinpoints could be seen and these were observed to a lesser extent in the yellow stones. In the greenish-yellow diamonds the pinpoints were isolated.

All three types contained metallic inclusions and some grainy structures could be seen on the surfaces of faceted stones. None of the diamonds showed absorption in the visible. Some of the greenish-yellow colour may arise from nickel playing a part in crystal development.

The De Beers synthetic diamonds have been classed as Type Ib from the dispersion of their nitrogen into single atoms. Most natural yellow diamonds are Type Ia and yellow Type Ib diamonds are rare in nature.

None of the De Beers diamonds was electroconductive, nor did they show a high degree of thermal conductivity; this property is shared by the GE and Sumitomo diamonds. The De Beers diamonds were weakly magnetic but this property alone does not distinguish them.

In 1993 De Beers grew some blue diamonds for research work, the blue arising from a boron dopant. GIA looked at three polished diamonds, coloured light bluish-greenish-grey, dark blue and near colourless, with clarity grades from VS1 down to SI (synthetic diamonds are not yet clarity graded commercially).

The light bluish stone fluoresced a weak orange under LWUV, with uneven colour distribution and a weak orange phosphorescence lasting from one to five minutes. The dark blue stone fluoresced slightly yellowish-orange of moderate intensity with a strong persistent phosphorescence of the same colour.

The near-colourless stone was inert under LWUV, with a slightly greenish-yellow response to SWUV, with a very strong yellow phosphorescence. None of these effects is shown in any natural diamond.

All three stones contained distinct internal growth sectors and similar pinpoint inclusions like those seen in other synthetic diamonds. Metallic flux was also present. Colour zoning was strong and strain birefringence could be seen between crossed polars.

The stones were found to be a mixture of Types IIa, IIb and Ib, a combination not found in any natural diamonds. General properties were in the diamond range and no visible absorption could be seen. As we have pointed out in *The Identification of Gemstones* the danger with all these synthetics is when they arrive for testing in very small sizes.

Russian Diamonds

As we have seen, the first synthetic diamonds were grown for research purposes rather than for use in ornamentation. Diamonds grown in Russia during the 1990s were intended for jewellery use. GIA tested five faceted stones with colours ranging from yellow to orange or brownish-yellow and three stones coloured yellow to greenish whose colours may have developed through treatment. The three stones may have been early examples of high-pressure/high-temperature (HP/HT) treatment, as the colours achieved are similar.

The non-treated diamonds showed a greenish-yellow fluorescence under LWUV with varying intensity but there was no phosphorescence. The treated stones fluoresced a very strong greenish-yellow under LWUV, with a moderate to strong yellow phosphorescence. Under SWUV the treated diamonds gave a stronger yellowish-green fluorescence with a moderate to strong yellow phosphorescence.

Under a hand spectroscope after the specimens had been cooled with a spray refrigerant some of the Russian synthetic diamonds showed a sharp absorption band at 658 nm with a sharp but weaker band at 637 nm in one specimen and at 527 nm in another. The treated stones were cooled in the same way and showed several absorption bands between 600 and 470 nm with less absorption below 450 nm.

When examined under a strong green light several of the samples showed uneven weak to moderately intense luminescence. Some natural diamonds of this colour have been reported to give a similar response. The Russian diamonds combined features of both Type Ia and Type Ib diamonds. Synthetic diamonds of the same yellow colour are more commonly pure Type Ib.

In summary the Russian synthetic diamonds were the first to show fluorescence under LWUV and the first in which heat-treated specimens fluoresced more strongly under LWUV than under SWUV. All the heat-treated stones gave a yellow phosphorescence, so we can say that a diamond which responds to either type of UV in any way needs careful investigation.

Absorption bands between about 658 and 637 nm and between 560 and 460 nm seem to have occurred so far in Russian synthetic diamonds only. In general the variability in the results obtained in testing these stones may indicate their experimental nature.

Isotopically Pure Carbon-12 Diamond

Isotopically pure carbon-12 diamonds for research purposes were reported by the makers, GE Research and Development Centre, Schenectady, New York, USA, in 1990. It is possible, as always, that examples will leak into the trade. The two stones reported by GIA are said to show some signs of strain birefringence, which can be seen in natural but not so far in other synthetic diamonds.

The stones also showed inclusions resembling metallic rods and clusters of minute triangular or lozenge-shaped tabular material beneath the octahedral crystal faces. Scattered pinpoint inclusions with a white, metallic appearance could be seen by reflected light and gave a brown appearance in transmitted light.

The crystals showed no graining but strain birefringence could be seen with a weak pattern of grey or blue. Neither specimen responded to LWUV but there was a weak yellowish-orange fluorescence under

SWUV. When the source was switched on the colour appeared to increase in intensity before settling to a fixed level.

A cathodoluminescence test was found to be useful as the specimens showed a zoned patterning correlated with different growth sectors. The colour was slightly greenish-blue. Under X-rays the stones fluoresced yellow with a very persistent yellow phosphorescence, which lasted for up to ten minutes in one crystal. Neither of the two specimens showed electroconductivity and neither showed absorption bands in the visible region of the spectrum. The crystals were assigned to Types IIa/IIb. The more recently grown GE diamond crystals fluoresced strongly under SWUV and contained metallic inclusions.

The reports above can also be found in O'Donoghue and Joyner, *The Identification of Gemstones*, where they accompany a text which deals equally with naturally occurring gemstones. Most of the full data can be found in different issues of GIA's *Gems & Gemology*. In the same text we summarized the characteristics of synthetic colourless and near-colourless diamonds.

Gemmologists should check a known diamond for response to SWUV and for the presence of metallic inclusions which can be present only from the growth process. They should also look for response to a rare earth magnet since many synthetic diamonds contain enough metallic inclusions for their response to be easy to detect. They should also examine the behaviour of synthetic diamonds between crossed polars. Since most diamonds will not show complete and even extinction but anomalous birefringence which will usually be due to strain, the effect seen between crossed polars may not in itself be enough to diagnose a synthetic diamond. A black cross when tested in this way suggests synthetic diamond.

Synthetic colourless or near-colourless diamonds will be type IIa or mixed Types IIa+Ib+IIb and will be colourless, light grey, very light blue, yellow or green with some of the colour in sectors. The presence of metallic inclusions or of clouds of pinpoint inclusions is strongly suggestive of synthetic diamond. Learn to distinguish between surface dust and very small included material.

Colourless to near-colourless synthetic diamond does not usually respond to LWUV but is likely to show fluorescence and phosphorescence under SWUV with the most common colours being yellow, greenish-yellow or orange-yellow. The effects may be weak or strong.

Most synthetic diamonds that are uneven or take a square, octagonal or cross-shaped pattern phosphoresce after SWUV irradiation, some specimens showing the effect persisting for up to one minute. The phosphorescence colour is most commonly yellow or greenish-yellow. Under visible light there is generally no fluorescent response. No sharp absorption bands can be seen.

Gemesis Diamond

Gemesis laboratory-created diamonds were extensively described in the Winter 2002 issue of *Gems & Gemology*. These are high-quality Type Ib diamond crystals grown by the Gemesis Corporation of Sarasota, Florida, and sold by them. Samples described weigh up to 3.5 ct and in some cases colour zoning with yellow and narrower colourless zones can be seen and a weak fluorescence under UV showing a small green cross-shaped zone combined with a weak overall orange fluorescence. These features serve to identify the crystals. Some metallic inclusions have also been reported. The paper states that the Gemesis diamond crystals can be identified either by EDXRF chemical analysis or by the De Beers DiamondView luminescence imaging system.

The company state their intention is to attain a level of production which will give 90 per cent of crystals suitable for cutting polished stones of 1 ct or above. There are no plans, the paper states, for the production of crystals which would yield polished melée up to 0.20 ct. Selected manufacturers and retailers have been targeted by the company in the initial sales drive. It has been proposed that colourless and blue diamond crystals will be attempted.

Growth and the apparatus for growing is described in the *Gems & Gemology* paper. Pressure is applied to a growth chamber measuring about 2.5 cm a side via sets of anvils in groups of eight outer and six inner anvils. Pressure to the anvils is applied hydraulically and the chamber is heated by a graphite element.

Typical diamond growth conditions with this apparatus are pressures of 5.0–6.5 gigapascals (about 50–65 kilobars) and temperatures of 1,350 to 1,800°C. Metals such as cobalt, iron and nickel give a solvent and catalyst medium.

Growth takes place on a very small diamond seed which may be either natural or synthetic diamond. The shape of the final grown crystal is determined by the orientation of the seed.

At the time the paper was written Gemesis was operating 23 BARS (the name given to the apparatus); with the equipment currently in use it is possible to grow a single diamond up to 3.5 ct in about eighty hours.

Some colourless and some blue diamond crystals had been produced but no commercial production had taken place at the time of writing.

GIA examined thirty Gemesis diamonds: five orange-yellow and yellow-orange crystals, weighing between 1.81 and 2.47 ct and twenty-five faceted specimens. Cryogenic techniques were used during spectroscopic studies.

The five crystals were nearly equant or slightly distorted from slight differences of growth conditions during a run of growth. The twenty-five faceted stones varied from yellow-orange through orange-yellow to yellow. The colour saturation was high and moderate in tone. Clarity was good, specimens being relatively free from inclusions compared to other examples of synthetic diamond. The equivalent clarity grade of the test samples would be in the VS to SI range. Ten of the faceted stones showed small metallic inclusions visible under 10x magnification and the majority contained cloud-like arrangements of very small pinpoint inclusions. Some stones retained remnants of the original crystal surface that polishing did not remove.

Colour zoning could be seen in most polished specimens under magnification. Examined through the pavilion, most polished diamonds showed larger yellow zones which could be seen to be separated by narrower colourless ones. In some examples this effect could be seen through the crown. In two of the polished diamonds no colourless zones could be detected. Careful positioning and shaping of the crown facets helps to reduce the visibility of the zones. Modifications to the growth process can also minimize visibility of the zoning.

None of the crystals showed any trace of a seed though a remnant of the imprint of the seed could be seen on their bases. In general the seed breaks away from the crystal on removal from the growth chamber.

Compared to other synthetic diamonds the Gemesis product shows a weaker luminescence under both forms of UV radiation. Ten of the specimens examined were found to be inert. The fluorescence colours seen in the remainder of the samples was a weak or very weak orange to both LWUV and SWUV. In the other nine samples the reaction to SWUV was either slightly weaker or slightly stronger in intensity as compared with LWUV. A small green cross-shaped pattern could be seen superimposed on the weak orange luminescence in some samples (ten in LWUV and fifteen in SWUV). In two of the crystals the cross shape could be seen near the centre of the base where the seed had been. Similar cross-shaped patterns but of greater intensity could be seen in all the samples examined by a Luminoscope cathodoluminescence unit and the De Beers DiamondView.

None of the samples phosphoresced from desk lamps or from UV sources.

Examination for visible spectra showed no sharp absorption bands in any of the samples tested, though there was a gradual increase in absorption towards the blue end of the spectrum.

Testing by infra-red spectroscopy showed that the diamonds were Type Ib, several specimens showing a weak absorption feature at 1,284 cm^{-1} (this was due to the aggregated 'A' form of nitrogen). The differences in colour hues could not be correlated with any significant difference in the visible spectra.

The spectra, with increasing absorption towards the ultra-violet, beginning around 500 nm, were characteristic of Type Ib diamonds, natural or synthetic.

The two yellow stones from which photoluminescence spectra (PL) were taken showed a single band at 1,332.5 cm^{-1}, which is the first-order and characteristic Raman peak for diamond with additional sharp peaks at 747 and 637 nm in one of the samples. Three out of the four samples analysed by EDXRF showed no X-ray emission peaks while one showed peaks ascribed to nickel and iron, these elements deriving from the solvent material used for growth by the temperature gradient technique. Other synthetic diamonds have shown visible metal inclusions and some natural diamonds contain iron as oxide or sulphide.

At the time of writing the presence of nickel can still be taken to be proof of artificial origin in diamond.

In the orange-yellow stones sharp or broad emission peaks were recorded from one or more of the six samples tested whose PL spectra were obtained. Many of the peaks are believed to be due to nickel-nitrogen, cobalt-nitrogen or nitrogen-vacancy complexes in the crystal structure. Figures are, 747, 701, 672, 637, 614, 581, 573, 566 and 543 nm.

Two of the seven samples examined by EDXRF showed no X-ray emission peaks while the other five specimens showed peaks due to nickel, iron and/or cobalt. The presence of nickel or cobalt is taken as proof that the specimen is synthetic.

In one or more of the five yellow-orange specimens sharp or broad peaks were noted in the PL spectra: numbers were 805, 794, 776, 753, 747, 727, 701, 694, 639, 637, 623, 604, 599, 581, 575, 547 nm.

Of the six samples tested by EDXRF three showed no emission peaks while the spectra of the other three showed nickel-, iron- or cobtalt-related peaks.

More general means of identification are set out in the *Gems & Gemology* paper. When describing yellow synthetic diamonds grown in Russia several authors characterized the crystals as having a cubo-octa-hedral shape with striated or dendritic crystal surface features. In addition colour zoning could be seen; this was related to significant crystallographic directions. Planar or angular graining is seen along growth sector boundaries and patterning of greenish-yellow or yellow luminescence of varying intensity corresponding to the patterns of the colour zoning.

Opaque, metallic, small black inclusions or triangular shape are also characteristic of the Russian-grown diamond crystals, as are minute pinpoint inclusions. Other distinguishing features can be seen by the use of advance instrumentation.

On the other hand the Gemesis diamonds show fewer features that a gemmologist would be able to distinguish without recourse to more advanced techniques. The yellow to colourless zoning, metallic inclusions and weak cross-shaped fluorescence patterns all point to artificial origin. The colour is more saturated than that seen in most natural and other synthetic Type Ib yellow diamonds. The cathodoluminescence and the luminescence images detectable by the De Beers DiamondView lumines-cence tester of the Gemesis diamonds are unique to synthetic diamonds.

IR and visible spectra did not aid identification. Photoluminescence spectra showed weak emission peaks arising from nickel or cobalt which indicate synthetic diamond. These elements can be detected by EDXRF spectra. The absence of iron or cobalt is not indicative of natural origin, however, since two of the yellow, two of the orange-yellow and three of the yellow-orange Gemesis diamonds did not show their presence.

The Gemesis diamond growth team have also produced some experimental diamonds. These include a colourless Type IIa 0.20 ct round brilliant showing no X-ray emission peaks in the EDXRF spectrum and only a weak fluorescence under SWUV but with a persistent greenish-blue phosphorescence (more than one minute) under the same radiation, a major identifying feature of colourless synthetic diamonds; a bluish-green irradiated Type IIa round brilliant of 0.42 ct with an X-ray emission peak due to iron and a GR1 radiation band at 741 nm in the absorption spectrum; a blue Type IIb crystal of 0.55 ct with X-ray emission peaks due to iron and cobalt; a green irradiated Type IaA 0.65 ct rectangular-cut and a yellow-green Type IaA 0.42 ct square-cut stone, both showing facet-related colour zoning characteristic of treated coloured diamonds irradiated by electrons or other charged particles. The 0.65 ct specimen also showed an X-ray emission peak due to iron.

Diamonds Grown by Chemical Vapour Deposition (CVD)

A major paper in the Winter 2003 issue of *Gems & Gemology* describes the growth of brown to grey and near colourless single crystal Type IIa synthetic diamonds by Apollo Diamond Inc. of Boston, Massachusetts, USA. The diamonds have gemmological properties which enable them to be distinguished from natural diamonds and also from HP/HT treated diamonds, though such tests can be carried out only with laboratory equipment not usually available to the gemmologist.

The diamond crystals are grown on a substrate which may be natural or synthetic diamond. The CVD technique was patented by Linares and Doering in 1999 and 2003 (the first patent is United States patent 6, 582, 513, filed 14 May 1999).

While the Apollo Company produced some brown-to-grey and near colourless gem-quality crystals and some faceted stones the growth of these diamonds was not intended to be for ornamental use alone. Wang et al. the authors of the paper in *Gems & Gemology*, review the progress of synthetic diamond manufacture since 1955 when the first specimens were announced. Since then HP/HT-grown diamonds (not the treated ones) have been found to show cuboctahedral form and recognizable growth sectors, indicating the type of growth. Most early synthetic diamonds are yellow.

The CVD process does not include high pressures for changing carbon into diamond. Instead a gas-phase chemical reaction deposits layers of a synthetic diamond film on a solid substrate. If the substrate is a natural or synthetic diamond a single crystal diamond is produced; if the substrate is not diamond a mainly polycrystalline diamond results.

A diamond thin film was produced before the announcement of the GE diamond in 1955 and developments in the technique were reported from time to time. The CVD method uses gaseous reagents, often methane in hydrogen in a chamber with a substrate. A reaction takes place at high temperatures and low pressures, with the reactants being transported through the chamber by diffusion and convection. A number of reactions takes place on the surface of the substrate and a continuous layer of synthetic diamond forms. It is now possible to grow a colourless, high-purity, single crystal layer of a few millimetres on a diamond substrate.

While this technique has been used to grow films which have been used to coat gemstones the practice has been found to be uneconomical.

At the time of writing, Apollo Diamond were proposing to begin production of CVD diamonds for use in jewellery. It was expected that brown to near colourless Type IIa crystals up to 1 ct or more would become available; in addition some colourless Type IIa and blue Type IIb were expected by 2005. A total production of 5,000 to 10,000 ct of synthetic diamond is planned in the first instance with some faceted stones reaching 1 ct by mid-2004.

Details of the process are given by Wang et al. They also describe the diamonds in terms which will both warn and assist gemmologists in recognizing CVD diamonds. Out of eight crystals they examined, five showed evidence of a substrate and in four of these the substrate was Type Ib synthetic diamond. The fifth specimen had a substrate of CVD synthetic diamond. The other three crystals seemed to have lost their substrate and without it were a transparent light brown or grey to near colourless.

The crystals examined were tabular in shape with two near-parallel surfaces. Additional small octahedral and dodecahedral faces were present on one of the crystals.

A diamond grown on a substrate of Type Ib HP/HT synthetic diamond had a slightly yellow appearance owing to the presence of isolated nitrogen impurities in the substrate. The CVD portion of this stone was light brown viewed from the side. The boundary between the two portions of the stone was sharp and could be seen with the unaided eye. Under UV (the De Beers DiamondView) the two portions of this stone showed different responses.

All the faceted stones described by Wang et al. had been cut to get the best yield from the tabular crystals. They showed an evenly distributed dark brown colour varying from faint to fancy dark brown; one stone was near colourless. This became fancy brownish-yellow after polishing – the heat of the polishing may have been responsible for the coloration.

No graining or colour zoning were observed when magnified through the table facet. Through the girdle, however, which is at right angles to the growth direction of the tabular crystals, one specimen at least showed more than five brown planes which recorded growth disruptions of various kinds. Banding in natural brown diamonds arises from plastic deformation during geological processes. The planes in the CVD diamonds are narrow and show a sharper boundary compared to the slip bands in natural brown diamonds.

The CVD stones varied in clarity between GIA grades VS_1 and SI_2 and only a few small inclusions were noted, which may have been non-diamond carbon; there were no flux inclusions, of course, since no flux is used in the growth process.

Anomalous birefringence was seen as cross-hatched bands of low interference colours looking through the table facet. The effect may be due to uneven deposition during growth.

Wang et al. found that the response to UV varied among all the specimens examined. Eight showed no response to LWUV and most of the samples gave a very weak orange, orange-yellow or yellow.

All but one specimen responded to SWUV with colours ranging from very weak to moderate in orange or orange-yellow. Stones with a HP/HT diamond substrate fluoresced more strongly with a chalky green-yellow.

When the CVD diamonds were subjected to high-energy UV in the DiamondView machine they responded with a strong orange-red fluorescence. This is very rare in natural Type IIa diamonds. It is possible that in some cases at least an HP/HT diamond substrate could be responsible.

The IR spectrum of the CVD diamonds was also examined; eight of the samples showed readings which suggest that trace amounts of nitrogen impurities were present, a feature of Type Ib diamond. Readers are reminded that all the CVD diamonds examined by Wang et al. were Type IIa.

For the gemmologist some of these points may useful: a brown colour, a polished stone with shallow depth, and characteristic strain pattern (anomalous birefringence). In the laboratory the strong orange-red response seen with the DiamondView is probably the best indication.

Diamond Grown by High-pressure High-temperature (HP/HT) Methods

In 1999 General Electric were reporting the processing of diamonds by HP/HT methods, the aim being to make them as colourless as possible. At that time an arrangement was made with Pegasus Overseas Ltd (POL) that the girdles of all the HP/HT treated diamonds would be inscribed GE POL. The name Belletaire is now used. GIA found that 99 per cent of the diamonds were Type IIa and were originally brown or brownish. Internal graining is notable and there are partially heated cleavages. While the lettering can be removed some persistent traces have been reported.

GE POL diamonds had a low concentration of single nitrogen compared to similar untreated diamonds. Low concentrations of single nitrogen were responsible for any yellow coloration.

The Fall 2002 issue of Gems & Gemology makes the point that the advent of HP/HT diamonds has made the recognition of inclusions even more critical and reduces the number of people able to distinguish altered from natural diamonds (and of course other species which have undergone this type of treatment). Two diamonds shown to GIA showed how important microscope examination now is. One was a 3.38 ct marquise graded fancy intense blue, the other a 1.20 ct fancy vivid orange-yellow grade. Both showed inclusions strongly suggestive of HP/HT treatment. The blue diamond showed a large, flat and apparently graphitized inclusion with a colourless halo. The yellow stone also had a large flat inclusion, apparently graphitized and with a colourless halo.

The halo in the blue stone was quite flat with little or no relief; it was very shiny in reflected light. The yellow diamond's halo appeared to undulate as it went around the dark centre and looked sugary with lower reflection.

Black inclusions are relatively common in Type IIb blue diamonds of natural colour, less so in naturally coloured yellow diamonds. Tension haloes can form naturally round inclusions in any diamond. HP/HT treatment can also create tension haloes around existing inclusions and at the same time initiate graphite formation along the crystal faces of the inclusions. This may give rise to a large, flat inclusion with a dark centre, surrounded by a transparent or translucent halo.

GIA points out, however, that there are differences between tension haloes from natural causes and those arising as a result of HP/HT treatment. The orange-yellow diamond was considered to have been colour-enhanced by HP/HT annealing while the blue one was considered to show natural colour. The sugary texture and the undulating appearance of the tension halo in the orange-yellow diamond suggested a rapid change in heat and pressure while the smoother appearance of the halo in the blue stone suggested a slower natural process. Spectroscopic examination proved that this supposition was correct in both cases.

In the Fall 2001 issue of *Gems & Gemology* GIA gives an update on blue and pink HP/HT annealed diamonds which had now taken the name Belletaire. The update was to a previous report in the same journal for Fall 2000. the specimens concerned were one blue and two pink faceted diamonds, the blue stone weighing 0.73 ct and the pink ones 4.57 and 7.54 ct.

The gemmological properties of all the stones were the same as for their untreated counterparts. Both the pink stones were Type IIa with a nominal amount of nitrogen and the blue diamond Type IIb with boron-related absorption peaks in the infra-red; it was electrically conductive.

On the 7.54 ct pink stone one pavilion facet showed an etched and pitted surface. This was believed to be due to the high temperatures used in the heat treatment and similar effects have been noted on other diamonds treated in this way.

With the Raman microspectrometer and 514 nm laser low-temperature photoluminescence (PL) spectra were obtained from each of the stones. These spectra are able to give great assistance in the determination of luminescence types in colour-enhanced diamonds. A paper in the same journal for Spring 2000 outlines spectroscopic evidence for treatment using GE POL HP/HT natural Type IIa diamonds.

The Spring 2003 issue of the same journal has descriptions of two Type IIa diamonds with defects related to their nitrogen content. One was an HP/HT annealed stone of 1.15 ct, carrying the colour grade of fancy intense green yellow; the other weighed 4.15 ct and had a natural colour and the grade fancy vivid pinkish orange. It is unusual for diamonds of this type to show nitrogen impurities (which are usually identified by the lack of appreciable absorption from 1,400 to 800cm[-1] in the mid infra-red range). The paper describes the IR spectra of both stones.

A paper in the Summer 2004 issue described synthetic coloured HP/HT diamonds (yellow, blue, pink and green) from an Asian source, which were by Chatham Created Gems of San Francisco. Some of the diamonds showed colour saturations very similar to those seen in naturally coloured diamonds. All specimens examined contained flux inclusions, although some were not easy to distinguish, forming sparse clouds whose appearance needed experience to recognize.

In general the diamonds did not differ appreciably from other HP/HT diamonds but the paper identified some interesting features. The majority of the irradiated pink and green diamonds had a very low nitrogen content – until then these had been hard to manufacture. It is possible that a low N-content was needed to obtain the lighter pink colours.

Other interesting points included the production of a green diamond combining both blue Type IIb and yellow Type Ib in the same crystal, giving a green face-up finished stone. Patterned colour zoning and UV response related to cuboctahedral morphology have been encountered in other HP/HT products, together with the presence of nickel, cobalt or iron in their composition. Blue and yellow colour zoning in the same specimen has not been recorded from natural diamonds.

Colours of Synthetic Diamond

SYNTHETIC BLUE DIAMONDS

These are Type IIb or mixed Type IIb+IIa. The colour can be anywhere in the range from light to dark blue, with some examples showing a greenish- or greyish-blue when small yellow growth sectors are present.

Internal growth sectors, some with a central octagonal shape surrounded by octahedral and cubic faces, are characteristic of many synthetic blue diamonds. Growth sectors show up very well under UV. Colour distribution is usually fairly even. Graining planes intersecting in patterns are often observed and hourglass shapes are common. A weak cross-shaped anomalous birefringence can usually be seen; opaque black metallic and clouds of minute pinpoint inclusions are often present.

As in the case of all synthetic diamonds there are no solid (non-metallic but mineral) inclusions. These are described in O'Donoghue & Joyner. Many natural diamonds do in fact show a wide variety of mineral inclusions, some of red and green colour.

Blue diamonds are inert to LWUV but under SWUV may show a yellow or greenish-yellow fluorescence. A stronger response to SWUV than to LWUV is characteristic of many synthetic diamonds but not of natural stones. The fluorescence is generally unevenly distributed and follows internal growth sectors with square or octagonal patterns. Sectors may show fluorescence or remain inert on a random basis. In most cases there is a moderate to strong yellow persistent phosphorescence lasting up to one minute. There is no fluorescent response to strong visible light and specimens show no characteristic absorption in the visible region. Blue synthetic diamonds may be attracted by a strong magnet and are electrically conductive.

SYNTHETIC YELLOW DIAMONDS

The most commonly grown synthetic diamond in the early days of synthesis, these may be Type Ib or Ib+Ia. If irradiated they will be Type Ib or IaA. If treated by HP/HT they will be Type IaA.

The range of colour may extend from orange-yellow to greenish-yellow or brownish-yellow. Type Ib or IaA treated stones give an orange to pink or sometimes a red colour after irradiation and heating to about 800°C while Type IaA treated stones heated from 1,700 to 2,100°C at high pressure show a yellow to greenish-yellow or yellow-green colour.

Growth sectors are usually visible and may be octagonal with additional cubic sectors. Colours are usually light and dark yellow. Synthetic yellow diamonds that have been heated and irradiated to give a pink or red colour, show pink or red zoning with some yellow sectors.

Graining planes can be seen between internal growth sectors with occasional intersecting of the planes to form patterns. Hourglass and other effects may sometimes be seen when the stone is viewed in transmitted light and between crossed polars a black cross interference effect has been reported.

Metallic flux inclusions and the whitish pinpoints seen in other colours of synthetic diamond are also present in the yellow stones.

Fluorescent and phosphorescent effects vary. In Type Ib synthetic diamonds there is no visible response to LWUV but Type Ib+Ia stones show weak to strong yellow or yellow-green fluorescence. Type Ib or IaA treated yellow stones give a strong green fluorescence with a weak orange phosphorescence in different growth sectors though in some stones only a weak orange is seen.

Type IaA treated diamonds give a notably strong greenish-yellow or yellow fluorescence. Under SWUV Type Ib stones show either yellow or yellow-green fluorescence of weak to moderate intensity while Type Ib or IaA stones give a weak to strong yellowish-green. Type Ib or IaA treated diamonds give a strong to very strong green with a weak orange fluorescence in different growth sectors. Some stones show only the orange fluorescence. Type IaA treated yellow diamonds give a strong greenish-yellow.

The fluorescent effects can vary in intensity with Type Ib or IaA stones showing them more strongly under SWUV than under LWUV. With Type Ib or IaA treated stones the response is of equal strength or stronger under SWUV. In Type IaA treated stones the intensity is stronger under LWUV

The fluorescence colours are evenly distributed, duplicating the growth sectors. Phosphorescence is uncommon in Type Ib or IaA diamonds but there can sometimes be a weak yellow or greenish-yellow persisting for some seconds. In Type Ib or IaA treated diamonds there may be a weak orange phosphorescence lasting for several seconds. In Type IaA treated diamonds there may be a strong persistent yellow phosphorescence lasting for up to one minute.

Visible light luminescence may be seen only in Type Ib and IaA diamonds with a weak to moderate green colour. Similar effects with a weak orange colour may be seen in other types. Type Ib diamonds do not respond to visible light.

In Type Ib diamonds no sharp absorption bands are usually seen but in cryogenic conditions a band at 658 nm may be visible. When cooled, Type Ib and IaA diamonds may show bands at 691, 671, 658, 649, 647, 637, 627 and 617 nm. In Type Ib or IaA treated stones several sharp bands may be seen at 658, 637, 617, 595, 575, 553, 527 and 503 nm. Type IaA treated stones may show sharp absorption bands at 553, 547, 527, 518, 511, 503, 481, 478 and 473 nm. They may be attracted by a strong magnet.

SYNTHETIC RED DIAMONDS

Two small (about 0.5 ct) dark brownish-red diamonds were examined by GIA and reported in 1993. They had entered the diamond trade. Both had pronounced colour zoning. Through the top of one of them square-shaped and superimposed cross-shape light yellow areas could be seen, surrounded by much larger red areas. The cross-shaped pattern could be seen almost directly under the table facet and looking through the pavilion facets the colour zoning could be seen in four positions round the girdle; it showed as light yellow zones surrounded by larger red areas. The other stone showed similar effects.

Under reflected light only one graining line could be seen on the table of one of them. On the other the table facet showed a graining pattern with some parallel polishing lines. Both stones were attracted by a pocket magnet, the cause being large metallic inclusions from the growth process.

Under both LWUV and SWUV the crown facets of one of the red diamonds showed unevenly distributed very intense green and moderately intense reddish-orange areas, with the green fluorescence corresponding to the light yellow zones already mentioned. These sections showed a phosphorescence of several seconds after irradiation by SWUV. The reddish-orange fluorescence was seen only at one small point near the girdle under LWUV but under SWUV this colour was seen on all the large dark red areas.

One of the stones showed a different, uneven luminescence pattern. Under either type of UV, looking through the crown, a very small area of red fluorescence could be seen near the centre of the table, with a narrow zone of green fluorescence surrounding it. Narrow bands of an orange fluorescence pointed from the green-fluorescing areas towards the four corners of the table facet where there were areas of stronger orange fluorescence. The remainder of the stone fluoresced a weaker orange-red and no phosphorescence was observed. When the other stone was illuminated by strong visible light a moderate green luminescence was seen in the yellow area. This effect was not seen in the other stone.

With the spectroscope some sharp absorption bands could be seen between 800 and 400 nm in the 0.55 ct stone. Some of the bands could be seen with the hand spectroscope, especially when the stone had been cooled: these were in the region 660–500 nm. In the 0.43 ct stone fewer bands could be seen but they were in similar positions. Both stones absorbed increasingly strongly towards the violet and had a broad absorption region from 640 to 550 nm.

Infra-red spectroscopy showed that the two stones contained elements of both Type Ia and Ib. Both stones showed fluorescence effects which had not been previously reported in other synthetic diamonds. The response to LWUV was particularly unusual. Nickel has been found in the stones, from the growth process, and in one of the stones the presence of particular absorption bands shows that it had been irradiated and heated.

Compared to natural diamonds several features distinguish these red stones. Some natural pink Type IIa diamonds show an orange fluorescence under UV but it has been reported by GIA that up to the time of writing no treated pink diamond had been described. Some red to pink natural diamonds, whose colour is known to be natural, show a blue fluorescence and have been classed as Type Ia. Some pink to red natural diamonds of Type Ib show absorption bands at 637, 595, and 575 nm which are signs of irradiation and heating.

Neither these stones nor two other natural treated diamonds of mixed Type Ib and Ia coloured pink to yellow and with a moderately strong orange fluorescence under LWUV and SWUV resembled the two diamonds under discussion since they showed no pattern of colour zoning and did not show the nickel absorption bands.

THE ENHANCEMENT OF DIAMOND

Before looking at the colour enhancement of diamond and why it is carried out it will be useful to remind ourselves that diamond's unique properties, folklore and appearance can sometimes combine to give the impression of geological rarity. A tour round Hatton Garden's shop windows will to some extent dispel this impression but there is a commercial problem when so many diamonds are white (colourless). Apart from their brilliance and fire (display of spectrum colours) diamonds cannot easily be distinguished from one another in quality unless there are prominent inclusions. For this reason a system of diamond grading has long been in place (there are several systems, in fact but the principles of each are the same).

The colour grading of diamond covers all shades, although fancy (strongly coloured) stones have their own schemes. The stones for which a colour grading system is most important are those which approach 'pure' white (although perceptions of this varies with each individual and at different times). Naturally the idea is to place the diamond you are selling into as high a colour grade as possible without making too blatant a claim. When you are buying the reverse will be the case.

Today most gemmological laboratories will issue certificates for diamond on which the colour and clarity grades are shown. One might expect the term 'blue-white' to feature at the top of the list of colour grades, as it occurs quite often in non-gemmological literature, but such labels are too easily misused. Today the top colour grade (from GIA) is 'D'.

Colour Enhancement

In general there is less likely to be any trouble with colour grades in the top classes than a little lower down. This is the area in which 'white' becomes more perceptibly tinged with yellow without the diamond appearing as a distinct and recognizable yellow. While details of the grading procedures are outside the scope of this book, it is well worth examining a few polished diamonds against master stones (graded by a laboratory). The stone to be graded is placed in succession next to a series of stones of colours ranging from white through off-white to perceptible yellow (no other colours are needed once you have got to yellow).

With practice it soon becomes apparent which of the master stones comes closest to your specimen and the grade is then confirmed. At first it is surprising how the faintest of yellow tinges shows up when a diamond which appears perfectly white on its own is examined with the master stones.

The grading systems have helped to reduce some of the practices used years ago to enhance the colour of a diamond. As yellow is the commonest unwanted colour several tricks are used to diminish its effect. A simple one is to show diamonds in a stone paper with a light blue liner. This makes the stone look whiter and the practice is used generally today both for diamonds and for other colourless stones.

More deceitful tricks (although they could hardly have deceived for very long) included the application of blue from an indelible pencil to the back facets but this would almost certainly be seen. The only chance of successful deceptions along these lines would be when diamonds were sold to an unsuspecting customer unused to buying diamonds. The legendary days of diamond selling in the doorways of Hatton Garden have long gone.

While diamond certificates have always carried weight and details of size, cut and colour, some also state the weight of the original rough. This shows the direction in which certification is going, with ever more details being required. In future, certificates for high-quality stones will also need to state whether or not the colour of the stone has been enhanced and this places additional responsibilities on gem-testing laboratories.

Whether or not colour enhancement will come to be accepted as a normal state of affairs in diamond is uncertain. As usual, skilled choice of name and promotion could work wonders (as with the colour term 'chartreuse' used for diamonds whose colour might once have directed them towards the 'industrial' classification. Brown diamonds have also been well advertised, the campaigns being helped by the discovery of one or two exceptionally large, fine specimens.

The use of radium salts to turn a diamond to a dark green colour was effective enough but the radioactivity induced in the specimens reached dangerous levels (recognized at the time and out of the question today). In any case, the green colour would not have been found attractive by the trade if the stones had ever travelled that far. Interestingly the stones thus treated are still radioactive. The chances of encountering such a diamond are remote in the extreme but there may be rogue operators around and the cooperation of a friendly hospital is all that is needed. Perhaps such stones may turn up from some of the more recently identified diamond-producing countries.

Just for the record, some diamonds treated by radium salts have been found to show spots on pavilion surfaces where radiation has been localized. Apart from the radioactivity, which is detectable of course by a Geiger counter, the colour is a much darker green than the green sometimes found in untreated diamonds. Polishing does not remove the radioactivity and the green is unaffected by heating.

Diamond's colour can be changed by more practical methods. Whatever the source, most treatment can be detected by the presence of an absorption band which extends from the infra-red to the visible yellow-green portion of the spectrum. This is known as the GR1 (general radiation) band and diamonds showing it will appear dark, the colour ranging from green to black. The sources (described briefly below) are not powerful enough to colour the diamond right through and any coloured diamond which, when viewed from the side, shows uneven colour or a colour extending only a short distance inside the stone from the surface, should be regarded with suspicion.

Some treated diamonds, when examined from below, show marks surrounding the culet which have been compared to the radiating spokes of an opened umbrella. These stones will almost certainly have been treated via the culet. Dark patches elsewhere usually show that the irradiation has taken place through the table. When such a stone is placed table down on a light surface the observer may detect a dark circle around the table.

Of the various forms of treatment, high-energy electrons generate heat in the specimen and the colour may be affected. The commonest colour developed from this process is a blue-green to green. It is not always easy to detect this treatment: in diamonds treated by electrons of lower energy the colour inclines to blue. While the GR1 band will normally be present it can be detected only by laboratory techniques not available to the dealer.

There has been a lingering doubt in the public at large about the possible radioactivity in gemstones believed to have been treated. In general the anxiety has been misplaced, though there have been isolated instances.

In the case of diamond, Nassau, in *Gems Made by Man*, describes a very dark green stone, which looked black, in which surface-reaching fractures contained metallic particles derived from the polishing process. The particles were found to be radioactive after the stone was irradiated. The stone was boiled in acid which removed the radioactivity though whether or not the particles were removed from the stone was not stated.

Nassau also reports the production of an unusual colour in a diamond which fluoresced greenish-yellow before irradiation, from which it was hoped a chartreuse colour would result. In the event, after neutron irradiation, an orange-red body colour with a bright orange-red fluorescence was produced. Some Type Ib properties may have been present. Nassau also describes another neutron-irradiated diamond: cloudy before treatment, the stone became a rich sky blue afterwards.

A natural Type IIb blue diamond will be a semiconductor and Nassau has wondered whether a darker blue could be obtained by irradiation. If this were possible, testing might be difficult since, if the first test was for electroconductivity, it would not be obvious that treatment had taken place.

When irradiation is carried out the colours produced from the (usually) Cape originals may still not be attractive enough for commerce. This is why many if not most irradiated diamonds are subsequently heated to give more acceptable colours.

On irradiation the commonest colour is a dark green or a dark greenish-blue. This can be heated to give first a lighter green followed by brown, then yellow as the temperature is raised.

After yellow has been reached, further heating will not improve or alter the colour: it will, in fact, tend to return to its colour before treatment.

Yellow diamonds are easily the most sought after ones resulting from the irradiation-heating process. When the stone is large and spectacular considerable effort has to be made by laboratories to ensure that the origin of the colour is established. In the 1960s the yellow Deepdene diamond, originally a Cape stone, was irradiated to give a deep green and heated to give a golden-yellow. Details of the arguments over this stone can be found in the literature – was the stone treated or not? Some years afterwards the scientist who actually carried out the irradiation told me that he had done so but by that time the true nature of the colour had been established. The stone weighs 104.52 ct.

Pink is sometimes produced by irradiation and heating and even red, probably the rarest of the natural colours of diamond. Any red or brownish-red diamond would be subjected to the most rigorous testing, and although some examples do turn up in the major auction house catalogues I cannot have noted more than three in the past thirty years (of course, I may have missed some). Diamonds that turn pink or red will be Type Ib. A strong orange fluorescence has been reported (this can also be found in natural pink diamonds). Three GIA-certified brownish-red diamonds were seen at a show a year or two ago.

When the temperature of heating exceeds 400°C the GR1 absorption band may be destroyed. Bands not seen in natural diamonds, at 595 nm (cited as 594 or 592 nm by some authorities) and at 503 and 497 nm may be developed. The result of the development of these bands is that the specimen will become coloured.

Clarity Enhancement

In *The Identification of Gemstones*, O'Donoghue and Joyner detail some of the first attempts to fill fractures and improve diamonds' clarity grade, made in the 1980s by Zvi Yehuda of Ramat Gan, Israel. It was reported that after cleaning, the stones were filled with a molten glass, the process being carried out at high temperatures. Glass remnants on the surface were removed.

Fracture filling has been carried out at least since the 1980s, and with stones as small as 0.02 ct and as large as 50 ct, according to publicity material. Even after more than twenty years, however, there are claims and counter-claims about the durability of the process, as well as discussions on disclosure.

The Israel-based firm of Koss & Schechter tried two different processes of filling diamonds in the 1990s. In both methods glasses were used, one a halogen glass, the other halogen oxide. At that time halogen glasses were in general use as fillings. Koss stones showed internal orange and yellow flashes when the diamonds were rotated under dark-field illumination. Under bright-field illumination the flashes were blue and violet. Not all filled diamonds showed these colours. All the filled areas contained gas bubbles and some diamonds showed flow structures in the fillings. Filling material was sometimes a distinct yellow but fine crackled lines seen in some other fillings by the same firm were not apparent in stones filled in the experiment under discussion. Lead and bromine were found in all the stones when examined by energy-dispersive X-ray fluorescence analysis.

Stones filled with halogen oxide glasses did not fill successfully. By 1996 a claim by the firm that they would incorporate a fluorescent additive in their filler had not yet been fulfilled and cathodoluminescence did not reveal the presence of any additives. A wide range of flash colours was observed in the diamonds that would not fill satisfactorily – the colours included red, orange, yellow, blue, purple and pink, seen under dark-field conditions. Under bright-field illumination the colours seen were bluish-green, green and greenish-yellow.

These facts were made known to the trade by Koss & Schechter in an unusual disclosure of working practices. The point was made that the RI of the filler should approximate that of the diamond host so that the filled area would be difficult to identify.

Stones belonging to and tested by GIA and which had been filled by Dialase Inc. of New York using a Yehuda-based filling process showed no sign of weight gain (the filling would be in the form of a thin film) while their appearance seemed to be enhanced in some cases and unchanged in others. Four out of the six stones tested showed a drop of one full colour grade after treatment and two stones showing no improvement in clarity also dropped to a lower colour grade.

Under magnification the filled diamonds showed a slightly greasy appearance with a slight tinge of yellow. The lens showed a large number of surface fracture signs, quite enough to alert a diamond dealer. The Dialase-filled diamonds showed the flash effect with a characteristic yellow-orange colour seen under dark-field lighting. Under bright-field lighting the same area changed to an intensely vivid electric blue; tilting the stone backwards and forwards showed the flash colours changing from orange to blue to orange. The effect is best seen when the stone is examined at a steep angle and close to a direction parallel to the plane of the treated fracture. The lighter the colour of the host the easier it is to detect the flash colours, which are hard to see, for example, in a dark brown diamond where only the blue colour will be apparent.

In very small fractures the flash effect may not be visible. Care should be taken not to mistake the orange flash for the iron-staining patches often seen in diamonds and other stones. Since the filled areas are in fact thin films they may show interference colours which will appear as multiple colours resembling rainbows, though not when the fractures have been filled. Flow structures and gas bubbles are common in glass-type fillings and if the filled area is carefully examined with the lens or microscope it may be seen to show a light-brown, light-yellow or orange-yellow colour. It has been suggested that the filling material used in the Dialase experiment is a lead-bismuth oxychloride with an RI close to that of diamond.

The Dialase filling was not affected by ultrasonic cleaning, steam treatment or boiling in a detergent solution. The filling was unharmed by thermal shock and by stress induced by the setting process. It was damaged by repolishing in some instances and when the flame of a small torch used in repair was brought close to the filling signs of sweating in the form of beads on the surface were apparent. The best-illustrated summary of these experiments can be found in *Gems & Gemology* Fall 1994. The paper makes the point that flash colours are the best clue to fracture filling. Readers can also refer to a most useful chart published by GIA, which indicates in colour the main features of filled diamonds, covering flash colours, trapped bubbles, flow structure, misleading features such as interference colours, feathery appearance in unfilled breaks, natural iron staining, brown radiation staining and burn marks on the diamond surface left by the polishing process (these can be mistaken for remnants of filler left on the surface). The chart accompanied *Gems & Gemology* for Summer 1995.

Diamond Thin Films

While thin films can be applied to any gem species, in the case of diamond they could be used to alter the colour of a polished stone. The technique to some extent parallels the blooming of camera lenses, now carried out as a matter of course, and the thin films used have a thickness of around 1µm (0.001 mm). The films are applied to the surface by CVD. The thin layer helps to resist abrasion and to improve the stone's appearance.

The presence of thin films can usually be spotted by interference colours on the surface (this is due to the difference in RI between film and host) and is particularly notable when in rare cases a film of air occurs between surface and coating. The films cause an unusual haziness best seen in dark-field illumination – this is due to light scattering within the film. When a coated stone is examined in diffused light and held against a white background the film causes the surface to appear brown. The coating has a polycrystalline composition so that it does not show extinction between crossed polars and it shows no absorption in the visible.

Thermal conductivity tests have given similar readings for the coating and a silicon substrate so that coating a non-diamond with this type of film would not deceive the tester. For a coated non-diamond to pass as diamond, the diamond thin film would have to be at least 5 μm thick.

The colour of a polished diamond might be altered by a thin film coating but the technique seems not to have proceeded beyond the experimental stage, in which a blue film of 20 μm was deposited on a natural near-colourless diamond octahedron, giving a blue electroconductive crystal. Immersion should reveal this kind of treatment since natural blue diamond has a notably patchy colour and the film would show sharp edges. It is possible that a film could be applied to a Cape diamond substrate and the yellow and blue cancel one another out to given an 'improved' stone.

Coating by blooming as in camera lenses has been used with Cape diamonds. The coating can be seen when the stone is rotated in a strong light (the bluish bloom will then be seen) and there is a spotty or granular appearance in the girdle region, with pitting. Some coatings can be dissolved in sulphuric acid when the yellowish colour of the stone may be seen beneath.

Diamond crystals have been burnt to oxidize the surface and produce a whitish appearance, so that a yellowish crystal may resemble a whitish one with a frosted surface. Rumour has it that a diamond polisher was deceived by a parcel of crystals of this type which when cut returned to their yellow colour.

NOTES FROM THE LITERATURE

Rough diamonds have been imitated by an ingenious assemblage reported from Brazil. Two specimens, weighing 23.57 and 8.71 ct, gave specific gravity readings of 3.56 and 5.89 respectively. The 8.71 ct stone was identified as cubic zirconia and the larger specimen as topaz. Raman spectroscopy confirmed both diagnoses. In the smaller specimen one spot showed blue under LWUV; it turned out to be a colourless fragment of diamond attached by glue to a notch in the surface. An unscrupulous dealer could (and apparently did) show the unwary customer a diamond reading with a diamond probe, since he knew where to place the tip.

A treated black diamond before and after treatment was briefly described in the Fall 2001 issue of *Gem & Gemology*. In a 1971 issue of the same journal an intensely flawed black diamond of 10 ct was described; this appearance was more or less the norm for black diamond. The usual way of colour treatment in black diamonds has been irradiation, which tends to give what is really a very dark green rather than actual black. Irradiated diamonds almost always give small green flashes either in the body colour or reflected from a fracture.

In the 2001 note, GIA described a new type of black diamond which did not show the mottled 'pepper and salt' appearance seen in most natural black diamonds nor the hints of green in the irradiated stones. The newcomers showed extensive fracturing and low clarity and, in addition, a black lining in most of the surface-reaching fractures. Raman spectroscopy showed that the material of this lining matched the pattern for graphite.

Graphite has been found in fractures and around mineral inclusions in untreated diamonds but GIA suspected that in this case the blackening of the fractures might have been done artificially. In the experiment a milky white diamond of 0.085 ct was heated with close control in a vacuum in the temperature range of 900–1,650°C for periods ranging from a few minutes to several hours. This high-temperature treatment produced a predominantly black appearance. Under magnification it could be seen that the colour was caused by graphite lining the fractures while the near colourless areas adjacent to the fractures were unchanged.

Graphitization was confined to areas near the surface of the surface-reaching fractures and cleavages (graphitization is random in natural black diamonds). It was not thought that the diamond type was a critical factor in the colour obtained by the treatment as in HP/HT treated stones. Experiments were continuing at the time of the publication of the piece, and further reports are expected.

The Spring 2003 issue of *Gems & Gemology* described an instance of a diamond's body colour being affected by fracture filling. Some years ago GIA described how a 10.88 ct light yellow diamond was coated with pink nail polish with the intention of substituting it for a natural pink diamond that had been stolen.

Fracture filling is not regularly used to alter colour (a regular supply of suitable stones cannot be relied upon). The stone described in *Gems & Gemology* weighed 0.20 ct and was cut as a pink round brilliant. A preliminary examination showed uneven distribution of colour; it appeared to be confined to the fractures which could be seen by the unaided eye, while the remainder of the stone was near-colourless.

Magnification showed that the pink colour was strictly confined to the fractures with areas of concentrated colour forming bead-like shapes, an effect associated with some clarity-enhanced diamonds. The stone could have been filled with a fluid which hardened after impregnation.

However, there appeared to have been no attempts to enhance the clarity of the diamond as the fractures could easily be detected after filling; the filler had a low refractive index which made it easier to see and was not, presumably, intended to echo the refractive index of the diamond host. For this reason it was concluded that the purpose of the filling was to change the colour of the diamond. GIA were not able to identify the substance used.

In the Spring 2003 issue of *Gems & Gemology* a novel variety of synthetic gem diamond was introduced. The LifeGem was reported to have been made from carbon from cremated cadavers. The human body is said to contain enough carbon for at least ten 1 ct diamonds to be synthesized by a process known as carbon curing. The carbon was purified and transformed into graphite, although the report stated that boron was not always entirely removed in the process. Undertakers who wanted to use this technique had to use the proprietary technology of LifeGem.

The diamond illustrated with the article weighed 0.23 ct and appeared to be a light blue (presumably from the boron content). The company has been involved with the growth of synthetic diamond for a number of years and it was reported that the process used Russian-designed presses, using the temperature gradient method of growth with a flux solution of iron alloys. The boron content classed the diamonds as Type IIb. Faceted diamonds up to 1 ct were being produced, the colour being light to medium blue. There is no means of identifying the 'original donor' of the diamond so far.

Simulants of diamond made from aluminium nitride and from aluminium nitride and silicon carbon alloys were the subject of United States Patent 6,048,813 of 11 April 2000, assigned to C.E. Hunter. The single crystals of pure aluminium nitride or of the alloy can be doped with gallium, cerium, gadolinium and samarium, which are cited in the patent. Crystals can be grown by more than one technique and the patent gives details of them as well as describing the fashioning process. No gemmological properties are given, however. The patent can be viewed at www.uspto.gov/patft/index.html.

A bicoloured synthetic diamond crystal of 0.33 ct was reported in the Summer 2003 issue of *Gems & Gemology*. The stone displayed both blue and yellow areas and was a distorted cubo-octahedral shape. Much of the blue appeared to be confined to cubic sectors of the crystal but apart from this there was no distinct colour distribution. Metallic inclusions could be seen under magnification and these probably accounted for the stone's attraction to a pocket magnet. A strong green fluorescence, confined to no particular area, was obtained under SWUV and was followed by a strong and persistent green phosphorescence lasting for more than two minutes. The blue sectors seemed to luminesce more strongly than the yellow ones and the crystal was inert to LWUV. The blue zones conducted electricity and the yellow zones did not.

The cause of colour was investigated with the aid of an FTIR spectrophotometer but it was found to be impossible to obtain IR spectra for each colour separately. This meant that the different areas showed only slight variations but boron-induced absorptions were identified. It appears that the blue and yellow colours are due to a combination of boron and single nitrogen defects. Fluctuations in growth temperature are known to produce colour changes in synthetic diamonds during growth.

The *Gemmology Queensland* issue of June 2003 warned readers that CZ was now appearing in a brown colour, perhaps with a brown diamond imitation in view and as an alexandrite red to green change of colour imitation. A black variety which did not change colour when subjected to high temperatures is also reported.

Among the dopants used with CZ is cerium, which has not been used, so far as I know, with other artificial gem materials. Lacking a true ruby red, CZ makes do with the orange-red given by cerium – this is a bright and attractive colour. CZ cannot be successfully doped with chromium so an emerald green cannot be achieved. Nickel gives a brown colour and before readers wonder why brown CZ should be considered worth producing, remember that brown diamonds (once classed as industrials) have been very much in fashion over recent years.

Lorne Stather and I tested a small collection of synthetic diamonds with a pocket (iron) magnet and an RE magnet. Fifty per cent of the sample responded conclusively to the RE magnet whose power was shown effectively when the specimens, floating on leaves on the surface of tap water, were pulled along. The iron magnet appeared to have little or no effect.

CHAPTER SIX

Ruby and Sapphire

R uby and the various colours of sapphire are both colour varieties
of the mineral corundum. In the previous chapter we found that
geologically speaking diamond is not one of the rarer minerals.
Fine colour and clarity in ruby is much harder to find and this means
that no elaborate systems of grading and selling are needed as we saw
in the case of the diamond.

*Verneuil ruby. Incompletely calcinated, so the stone contains impurities within,
between the curved layers*

HISTORICAL DEVELOPMENTS

It is easy from most existing texts and records to assume that Verneuil was the first person ever to grow crystals of corundum and that the flame-fusion process is well-known to all gemmologists. This is probably true in the sense that his development of the flame-fusion process was the spur to its widespread application. We should remember, however, that corundum has other desirable properties, in particular hardness, which make its synthesis most desirable.

Verneuil's Predecessors

During the nineteenth century Gay-Lussac, Rose and Gaudin worked on the synthesis of ruby (Gaudin actually succeeded although he believed his product was glass). The crystals obtained by Gaudin were marked by cloudiness and a tendency to cracking. The melting point of ruby, at 2,050°C, was always, in the early years of research, a barrier to satisfactory synthesis but was overcome in the end by the development of the oxy-hydrogen flame.

In 1877 the French chemist Edmond Frémy published, with his co-

Frémy ruby. Typical triangular inclusions – the so-called 'coat-hangers'

worker Feil, a paper which showed that small but clear ruby crystals could be obtained. After Feil's death in 1876 Verneuil joined Frémy. Their work had employed ceramic crucibles through which humid air could diffuse and the recrystallization of alumina with potassium dichromate to provide the chromium needed for the colour.

In 1891 Frémy published *Synthèse du rubis*, with unusual photographs – in black and white overprinted in red – of growth methods and apparatus. The rubies shown were not boules of the familiar Verneuil flame-fusion type but small transparent platy crystals of rhombohedral form (Frémy's work on rubies with Feil had produced thin tabular crystals). The crystals depicted could be faceted into small stones and some of the illustrations show pieces of jewellery set with some of the rubies so fashioned. Examples of the apparatus and of the crystals are usually on view at the Muséum d'Histoire Naturelle in Paris.

The corundum crystals obtained by Frémy and Verneuil measured up to 3 mm and weighed up to 0.3 ct. Colourless crystals as well as ruby, violet and blue – and some both red and blue – were grown. It is interesting to note that on a tour of crystal growth facilities in Japan in the 1980s I was shown two crystals half ruby and half blue sapphire. It took some persuasion to obtain a specimen which must, like those of Frémy and Verneuil, have become coloured in this way by accident.

The Geneva Ruby

There were difficulties in achieving crystals of significant size and a few if any were produced in bulk. Contemporary with Frémy and Verneuil's work was the appearance in the trade of rubies which were reported to have been 'reconstructed' – produced from the reconstitution of molten fragments left over from cutting natural ruby. As specimens were sold as natural rubies by a Geneva firm they became known as 'Geneva rubies' and the name has stuck.

Examination of these stones showed the rounded gas bubbles now familiar from the boules of the Verneuil process. Disputes arose over nomenclature, at least in France, and specimens had to be described as artificial. Nassau, in *Gems Made by Man*, reports that Tiffany's gem expert Kunz, on examining the rubies, decided they were artificial and that they were grown by the process devised by Frémy and Feil. In the rough state the Geneva rubies were characteristically shoe-button-shaped: they had a convex upper surface with an irregular base. Nassau was able to work out how the crystals must have been grown; some aspects of the growth were similar to the later Verneuil boule manufacture.

The Geneva rubies lasted for about twenty years until the later Verneuil method was developed. Nassau experimented with the reconstitution of melted natural rubies and found that the resulting product showed as ruby pieces joined together by a polycrystalline dark grey and opaque corundum. Air bubbles were large and prominent. This experiment showed that the 'reconstructed' theory could not be substantiated. His stones were unlike the later Verneuil product.

SYNTHETIC RUBY

The Verneuil (Flame-fusion) Ruby

The first synthetic ruby in gem form appeared in 1902. It had taken Verneuil many years to arrive at successful synthesis. The starting material had to be perfected, and while chromium was needed for the red colour, iron has to be excluded to prevent a brown tinge. Ammonium alum and chromium alum were purified and mixed, then heated to provide the required mixture and to drive off water and ammonium and sulphur compounds. The resulting powder formed the starting material for the growth itself. This process is followed today.

The powder was fused in an oxy-hydrogen flame after falling vertically from a container on the top of the apparatus, the molten drops forming a single crystal on a rotating and descending ceramic pedestal. This single crystal became known as a boule.

For the gemmologist the orientation of the boule is important. Each boule is a single crystal rather than a polycrystalline aggregate. The boule's vertical axis corresponds with the vertical axis of the natural crystal. In corundum this is a direction of single refraction, so that stones cut from the boule with their table facets at right angles to this direction show the phenomenon of dichroism through their table facet. This can be seen only with a dichroscope, which presents two rectangular windows side-by-side.

When a Verneuil ruby is examined at right angles to the optic axis direction, one aperture will show an orange-red and the other a yellowish-red. Natural ruby crystals provide the best yield when the table facet is placed at right angles to the optic axis direction (the view through the table follows the optic axis) and only a single colour will be seen with the dichroscope.

Examination of the stones cut from the boules under the microscope shows two features at least which differentiate flame-fusion-grown materials from natural rubies. Large, rounded bubbles with notably bold edges are not seen in natural rubies (or other natural crystalline species), nor are curved colour and growth banding. The growth banding shows up as curved lines reminiscent of the grooves on a vinyl record: colour banding is also curved but much less precise. It is worth mentioning that the curved structures are not always as easy to see as textbooks have suggested – sometimes gemmologists find it easier to immerse the specimen in a liquid of similar refractive index (e.g. di-iodomethane) and use a green colour filter above the specimen and below the objective (lower) lens of the microscope.

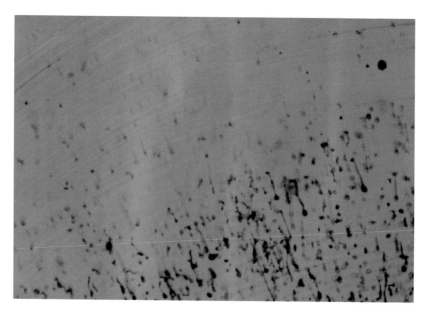

Verneuil ruby. Curved striae and a swarm of gas bubbles

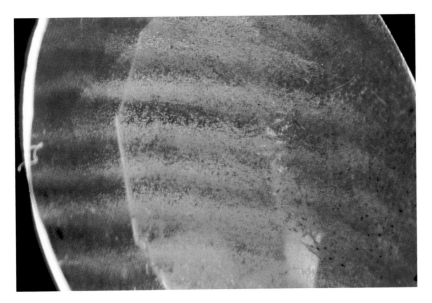

Verneuil ruby. Broad curved zones of tiny gas bubbles

Since natural ruby should also be examined first under magnification, the gemmologist should soon develop a familiarity with some of the inclusion patterns involving other mineral species (or, sometimes, corundum itself). The immediate impression when a Verneuil product is examined will be, 'How clean this is.' Natural inclusions will be absent. There will be no hexagonal crystals of apatite, no calcite rhombs nor small octahedral of spinel. Any growth or colour zoning seen will be curved rather than angular and the gas bubbles should be seen quite easily.

The characteristic liquid inclusions so common in natural ruby have long been known as fingerprints or feathers and appear flat under magnification. No liquid inclusions appear in flame-fusion stones, though veils of undigested flux in rubies grown by flux growth may resemble congregations of liquid droplets. From time to time such assemblies have been induced into Verneuil stones.

Both natural and Verneuil synthetic rubies may show asterism, the star effect. In corundum the star is most commonly six-rayed and in many specimens the centre may not be placed at the centre of the cabochon-cut stone. Any ruby showing a very prominent star, with evenly spaced rays with their origin exactly in the centre, should be examined for significant signs of flame-fusion growth, including rounded gas bubbles round the edges.

If, in addition, the stone has a polished flat back it is almost certain to be a Verneuil product. Star rubies have rarely been made by the other methods of ruby growth, though we shall see later that the Linde Company did produce some attractive star rubies in the 1960s.

Even with the Verneuil product whose cheapness might make it seem easy to spot, some ingenious activities have come to light. One is the sawing of parts of boules to make them resemble natural ruby crystals. Most, though not all, natural rubies are found as flat (tabular) crystals

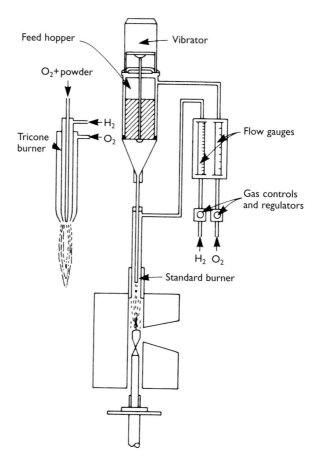

Fig. 7 *Diagram of the Verneuil furnace. The tricone burner is used for titanium synthetic stones*

of roughly hexagonal outline. Careful examination of the natural material will show signs that the underlying symmetry is three-fold rather than six-fold: upper surfaces show triangular markings and while small faces occur six times around the sides of the complete hexagonal crystal, they appear three times at the junction of upper surface and side, three times at the junction of lower surface and side.

Readers may wonder at this detail but we should remember that ruby crystals and the finished stones cut from them are of an altogether exceptional importance in the gem world. It is worth while, therefore, for cutters to fashion boule sections in the shape of natural ruby crystals, even to the extent of adding surface markings. I am not saying that this is a common practice but there are many gem and mineral shows, especially in the United States and Europe, and it is probably worthwhile trying out these crystals – then quickly disappearing. At the major gem and mineral shows many of those attending have considerable knowledge and experience of mineral crystals. Fewer will have seen what fine gem-quality crystals look like.

While rubies grown by the Verneuil flame-fusion method show no natural solid inclusions and the fingerprint-like liquid inclusions characteristic of natural ruby are also absent, rubies grown by other methods may show inclusions which, while not quite the same as those seen in natural stones, none the less could be mistaken for them at first glance.

Gemmological testing methods, apart from the microscope or lens, will not invariably enable you to differentiate between synthetic and natural stones, although the experienced eye may notice a stronger absorption spectrum (with the two close lines in the blue particularly prominent). Refractive index, birefringence and specific gravity are the same as for the natural material. The usual refractive index is near 1.76–1.77 with DR 0.009 and SG 3.99–4.00. The absorption spectrum is characteristic for chromium with an emission (coloured)/absorption (dark) doublet in the red, with two or three lines in the orange to yellow, general absorption of the green and three lines in the blue, two close together, and a general absorption of the violet.

Verneuil rubies can often be detected when a specimen is placed between crossed polars. In one direction, systems of straight lines can be seen, each system crossing the others at 60° or 120°. For this effect to be seen careful positioning of the specimen is needed but it can even

Verneuil ruby. Curved striae with discrete spherical gas bubbles

be seen in some colourless corundum. The lines are known as Plato lines and at the time of writing are diagnostic for synthetic corundum.

The crystal grower has control over what elements are present in his product. While most natural rubies contain some iron, concentrations are higher in stones from some localities. Thai rubies with a higher iron content than Burma ones are today heated to bring their colour up to 'Burma' standard.

When rubies are grown artificially, iron can be excluded. As iron inhibits fluorescence (which enhances the colour of the stone) the colour of the synthetic ruby is often arresting. When synthetic rubies are tested under LWUV they give a stronger red fluorescence than would be shown by many natural stones. While the interpretation of fluorescent effects is not completely reliable, a notably strong response should warn the gemmologist that additional tests should be made. This warning applies to all types of artificial ruby.

In the late 1960s the firm of Djévahirdjan in the Valais, Switzerland, produced the following list of the colours then available in their Verneuil-grown boules (the abbreviation sp indicates homogeneous colour inside the boule):

1	ruby: topaz light rose
1A	ruby: topaz rose
1 dark	ruby: topaz rose
1¼	ruby: topaz rose
1 bis	ruby: topaz rose
2	ruby: topaz dark rose
3	ruby: light rose
4	ruby: rose
5	ruby: dark rose
6	ruby: garnet colour (light)
8 sp	ruby: dark red or dark garnet
8	ruby: dark red
12	sapphire: white
20 sp	sapphire: lemon yellow
21 sp	sapphire: gold yellow
22 sp	sapphire: orange yellow
25 sp	sapphire: topaz of Brazil
30	sapphire: bluish of India
31	sapphire: Ceylon light blue
32	sapphire: Ceylon dark blue
33	sapphire: Kashmir blue
34	sapphire: Burma blue
35	sapphire: Burma dark blue
44	alexandrite: light for big stones
45	alexandrite:
46	alexandrite: dark
47	alexandrite: greenish
50 sp	danburite
55 sp	padparadschah
61	kunzite
65	sapphire: 'pourpre'
75	corundum: Djéva green

Rubies Grown by the Flux-melt Method

While the Verneuil flame-fusion rubies do well for some jewellery use, to the trained eye their suspiciously clear appearance (from lack of natural solid inclusions) makes them appear glassy. An alternative method of growth, employing a flux (solvent), produces a more realistic imitation of natural ruby. The crystals take up to one year to grow (Verneuil boules take an hour or two). Growth, and more importantly steady cooling, is computer-controlled and takes place in a precious metal crucible. All this makes for a much more expensive final product – while a 1 ct Verneuil, ruby may be bought for a few pounds, a flux-grown stone of similar size may cost over £100 or more.

Contrary to what is sometimes thought, flux growth was not devised for rubies for laser applications: traces of residual flux in the finished crystals give an unacceptable diminution of purity (ruby laser crystals need to be capable of fluorescence and flux traces would militate against that). The method of crystal pulling, described later, is far more use when ruby crystals of laser quality are needed.

The flux is a compound or compounds which, when melted themselves, will dissolve the starting material for corundum, whose melting point is near 2,050°C. The choice of flux is critical as it has an effect on the final shape of the crystal, which is often grown on a prepared seed, depending upon the intended application. Some fluxes give a flat, bladed crystal while others produce a more blocky shape. The latter is clearly of more use if a gemstone is to be the final product.

Flux-grown rubies contain no natural solid inclusions and no liquid inclusions. However, traces of flux in droplet form can assemble in structures which look like the liquid feathers/fingerprints characteristic of natural corundum. These structures can be deceptive when examined under magnification. The flux assemblies, however, are almost always twisted and can resemble net curtains or veils blowing in the wind. The same structures can be seen in flux-grown emerald.

Isolated fragments of flux have a metallic appearance under reflected light and the twisted veil-like structures can also appear metallic. Stones are best examined by indirect lighting, either by dark-field illumination or by a stand-alone fibre-optic lamp whose light can be manoeuvred as required. This in fact is a very good way of observing inclusions. It is important to note that the Plato lines seen in synthetic Verneuil flame-fusion ruby are not seen in specimens grown by the flux-melt or other methods.

As we have seen, flux growth uses a precious metal crucible. Fragments from the crucible wall may break off and become included in the growing crystal, and if they are not detected and cut away they can finish up in the fashioned gemstone. Under magnification these fragments appear as angular particles with a metallic lustre when light is reflected from them. They are diagnostic for this type of artificial ruby as no crucible is involved in flame-fusion growth.

The flux-grown ruby is a much more recent arrival on the synthetic gemstone scene than the Verneuil product. Patents for ruby growth by this method began to be published after the Second World War and growth by several manufacturers has gone on ever since. The best-known flux-grown rubies are those produced by Chatham and the Kashan and Ramaura products. Rubies grown by the late Professor P.O. Knischka at the University of Steyr, Austria, are remarkable for their beautiful crystals, which have been sold without fashioning into gemstones. These crystals have been reported to show faces not so far found in natural ruby crystals and are highly collectable for that reason and for their considerable beauty. Prices are very high.

Chatham rubies have been made as platy crystal groups with individual crystals larger than those in blue sapphire groups from the same manufacturer. Angular metallic platelets detached from the crucible have been reported. Single crystals are equidimensional rather than platy, suggesting that the flux was a tungstate or molybdate.

Chatham ruby. Fingerprint inclusion, highly reminiscent of liquid feathers in natural ruby

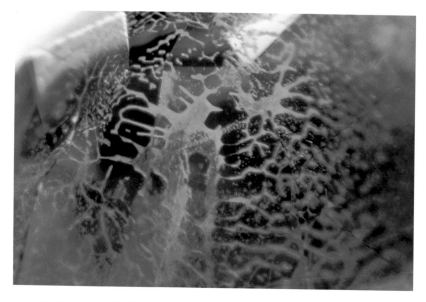

Chatham ruby. Highly characteristic, lace-like 'feather' (fingerprint inclusion)

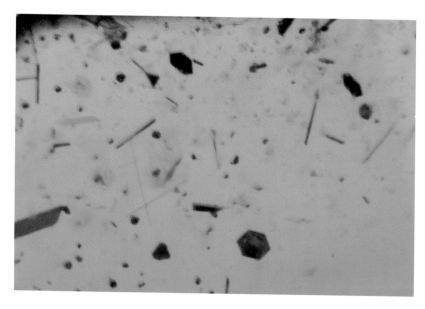

Disseminated black platinum platelets

Chatham ruby. Typical liquid 'feather' (fingerprint inclusion), easily confused with natural ruby

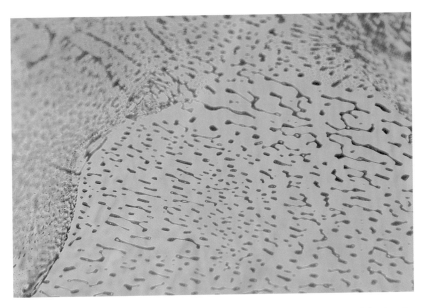

Chatham ruby. Variously shaped individual drops and different patterns of 'feathers'

KASHAN RUBY

While new flux-grown rubies appear on the market from time to time, some have held their own for at least forty years. The Kashan ruby was first grown by the firm of Ardon Associates Inc. of Dallas, Texas, in the 1960s. The firm sold not only faceted stones but crystal groups and single crystals. These were unlike the crystals of natural ruby; many showed a flat, platy habit, while natural ruby occurs as isolated crystals rather than in groups. None the less the Kashan stones seemed to alarm the trade at the time to the extent that the firm offered test sets for gemmologists not only to facilitate differentiation from natural ruby but also to enable some system of grading to be introduced for the artificial product.

The normal gemmological tests will not distinguish the Kashan ruby from the natural material. A microscope, however, shows an interior lacking natural solid inclusions but showing traces of flux which often resemble paint-splashes or breadcrumbs. In common with all flux-grown products there are no liquid inclusions. Some early examples showed flux particles in a moccasin shape and many contain the twisted flux veils seen also in other flux-grown rubies.

We have already mentioned the often startlingly strong red fluorescence shown by most synthetic rubies from which iron has been excluded. At one time it was reported that some Kashan stones contained additional iron aimed at minimizing this effect. The addition of iron would also prevent the transmission of SWUV. Some Kashan rubies do not in fact transmit it while most other synthetic rubies do. Iron may show bands in the absorption spectrum which would otherwise be characteristic of chromium alone.

Reports on what crystals growers are doing or have done abound, and it is vital to keep up with the literature so that unusual features do not go unrecognized. With the Kashan as well as in other cases there are always exceptions to what may seem to be a general rule. A good example is the heating and subsequent quenching in liquid of a Verneuil-type ruby with the consequent development of a crackled structure. The idea may have been to simulate the liquid feathers of a natural ruby but also may have been to simulate the flux veils of the much more expensive flux-grown synthetic ruby. It is always wrong to assume that crystal growers (or later handlers of the material) 'would never do that'. In the case described the gas bubbles would still have been prominent and identification – if suspicion was aroused – quite easy.

Kashan ruby. Conclusive flux inclusion consisting of cryolite

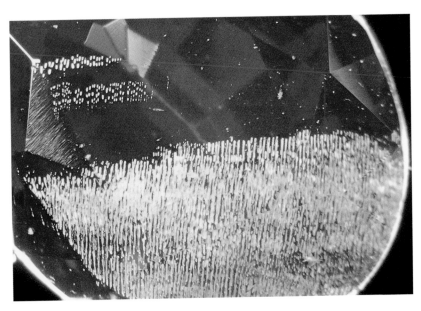

*Kashan ruby. 'Hoses' and flux-drop inclusions, in parallel alignment
and forming zones*

Kashan ruby. Gossamer-fine veils of flux inclusions

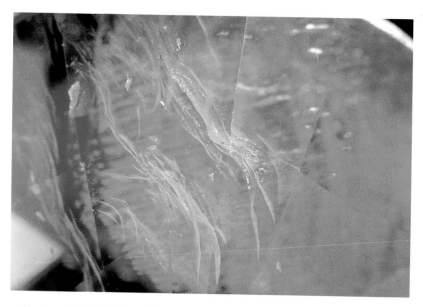

Kashan ruby. Typical fog and cloud-like inclusions affecting the clarity of the stone

CHATHAM RUBY

The flux method of ruby growth is also used by the San Francisco firm of Chatham, who have also grown different colours of sapphire. Their products can be identified as synthetic by clues similar to those seen in the Kashan rubies.

RAMAURA RUBY

In the early 1980s the Ramaura Division of Overland Gems Inc., Los Angeles, which later became the J.O. Crystal Company, began to produce high-quality ruby crystals under the name Ramaura Ruby. The published aim of the company was to sell faceted stones and lower-quality material for cutting into cabochons. High-temperature flux growth with spontaneous nucleation (i.e. not grown on a prepared seed) was said to be the method used. Seed growth allows the grower to keep closer control of growth rate and perfection while spontaneous nucleation, which may initiate growth from any site in the crucible, may give a variety of crystal habits and crystals which may be less included.

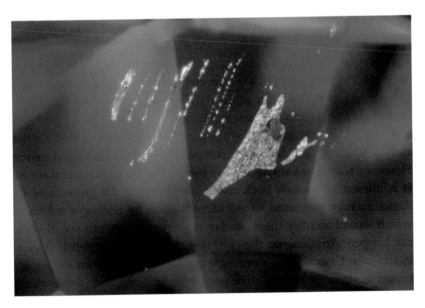

Yellow La-doped flux residue in synthetic Ramaura ruby (by J. Osmer)

As with other synthetic rubies, the colour of the Ramaura products does not distinguish them from the natural material nor from other synthetic ones. In the absence of a seed and consequent diminution of inclusion content the stones appear notably clear. In some cases tilting the stone may show up colour zoning and any inclusions present. Faceted stones do not always show dichroic effects in the same direction so crystals are presumably cut in directions giving the greatest possible yield. In any sample of crystals grown after spontaneous nucleation a variety of crystallographic orientations will be found.

On the refractometer the Ramaura stones do not show readings to distinguish them from natural or other synthetic rubies. There is no marked difference in specific gravity or in other properties. No phosphorescence could be seen under X-rays, something which is quite prominent in Verneuil-grown rubies. SWUV is not so effectively transmitted as in other synthetic rubies and while portions of some specimens have been reported to fluoresce chalky bluish-white or chalky yellow, giving a useful guide to Ramaura ruby, the areas giving these effects may be polished away by the lapidary.

Examination of the Ramaura stones under magnification has shown flux inclusions, some large, some coloured orange-yellow and some colourless. Some flux inclusions are grouped in such a way as to suggest the fingerprints in natural ruby. Angular metallic fragments of crucible material have not been noted.

Ramaura rubies can be both large and beautiful so that comparison with natural ruby is especially important. Particular attention needs to be paid to growth and any other visible lines in a specimen and their behaviour noted as the microscope focus is altered.

Some inclusions are apparently common to Ramaura and Kashan rubies. 'Comet-tail' structures composed of flux are characteristic. However, some similar effects have been reported from natural stones. It is interesting to look back and remember that the growers of the Ramaura ruby promised to aid identification by adding a substance which would identify them. As far as gemmologists can see, this did not happen. Anything on the crystal surface would be polished away in any case. Crystal groups or single crystals have not so far shown any unusual signs (which would probably be untypical fluorescence, though see above). The J.O. Crystal Company is reported to have ceased production in 2002.

DOUROS RUBY

In 1994 rubies grown in Greece by the Douros brothers appeared on the market, although at the time of writing they do not seem to be around in any quantity. Apparently two furnaces are used in growing the ruby crystals and both growth and cooling are slow. Cut stones up to 8.5 ct have been reported and a crystal of 350 ct was recorded in 1994. Some crystals are tabular and some rhombohedral, the latter being well suited for cutting. The crystals resemble the Ramaura product, which has also displayed penetration twinning, a phenomenon not reported from other artificial rubies. The colour of the Douros rubies ranges from a purplish-red or reddish-purple to a saturated red. The faceted stones that I have seen early on in the production were quite a dark red. It appears that more than one colouring element has been added and this assumption is probably correct since colour zoning with geometrical boundaries has been reported. Purple to bluish-purple colour zones in the form of intersecting acute-angled triangles have been reported with various other zone shapes; if the blue colour represents a sapphire component then both iron and titanium must have been added as colouring elements since both elements are required to produce the blue in sapphire.

Gemmologists will find that the Douros stones show normal physical and optical constants for ruby. Under LWUV some faceted stones give an intense orange-red fluorescence with a less pronounced red under SWUV. Under both types of UV some crystals have been found to be inert in their upper sections but this layer is usually lost when the crystals are polished.

Growth planes parallel to the principal crystal faces can be seen under magnification (an effect also noted from some Chatham and Ramaura rubies). In the Douros stones differently coloured areas, sharply defined, are notable: very deep red to near-colourless to light red are the principal colours reported. The sharp definition of the coloured areas is one of the best indications of the Douros product.

In general the sharply defined coloured areas found in both tabular and rhombohedral crystals, taken in conjunction with residual flux inclusions, either single or as droplets, are highly characteristic of this ruby. Some of the larger flux inclusions are in the form of rounded to elongated cavities containing a yellowish substance with bubbles or voids. Crazed patterning has been seen in the larger yellowish flux remnants. At the time of writing crucible remnants have not been reported.

GIA summarized the internal properties of the Douros ruby as resembling borax-treated natural ruby (which is discussed below). Rutile appears to be absent, which, in common with all non-Verneuil-grown rubies, rules out the production of stars.

LECHLEITNER RUBIES

The Austrian crystal grower Johann Lechleitner is well known for his emerald overgrowth on colourless beryl but he has also produced a complete ruby which has been sold both as faceted stones and as crystals. This activity was reported by Lechleitner to GIA in 1985. One ruby of 0.47 ct was a transparent saturated purplish-red with a slightly hazy appearance, while dichroism could be seen through the table facet (unlike in most natural rubies). Other properties are in the natural/synthetic ruby range and only the flux inclusions show that the ruby is artificial. Some specimens show Verneuil-like curved growth lines.

It was considered possible that growth took place on a Verneuil-grown seed, as this would account for the curved growth lines. It is also possible that a larger Verneuil crystal could have been doped with chromium to give the ruby colour and then acted as the seed itself upon which flux-grown ruby was deposited. A crystal grown in this dual way is in the collections of the Natural History Museum, London.

Growth by Crystal Pulling (The Czochralski Method)

The growth in the development and use of lasers in the 1960s and onwards led to the availability of crystals containing defects or other features which made them unsuitable for high technology use but which did not affect their ornamental potential. Though most crystals grown by pulling are virtually inclusion-free, there are not very many pulled rubies and fewer sapphires around, compared with the larger number of flux-grown and the far larger number of Verneuil-grown crystals. Gemmologists should regard the absence of natural solid inclusions as suspicious. The reason for the comparative lack of blue, green or yellow pulled sapphires is, naturally, the fact that iron is needed for their coloration. For any kind of laser application the presence of the fluorescence-poisoning iron is completely ruled out so that it would not be worthwhile for crystal growers to grow these materials apart from experimentally.

The pulling method can be simply described. An open crucible, made usually of iridium (which will not form unwanted compounds with the substance being grown, and which has a high melting point of 2,442°C) contains the melt, to the surface of which a seed crystal is

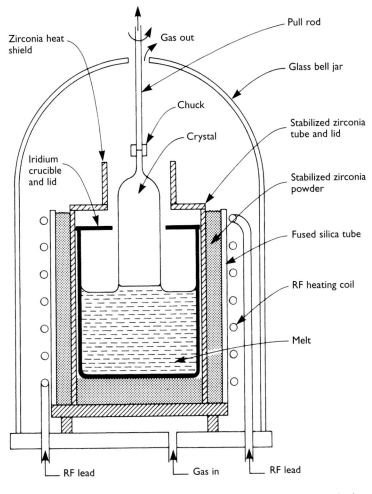

Fig. 8 Growing from the melt. This apparatus is used in the Czochralski method. The crucible can measure 6 inches (15 cm) across

lowered in such a way that it can, when raised, pull the melt after it. The temperature needed is critical as the seed crystal should not be allowed to melt nor to assist the melt to solidify if the temperature is too low. Pulling allows a rod-like crystal to be withdrawn from the melt, and is monitored and controlled by computer. There will be no solid inclusions, growth or colour banding and only the occasional elongated bubble may give a clue as to origin. The rate of pulling may be between 6 and 25 mm per hour, according to Nassau, *Gems Made by Man*.

Though Czochralski-grown ruby crystals were the only type considered suitable for laser applications twenty to thirty years ago, today the Verneuil technique, with suitable modifications to the method of heating and close computer control, can produce discs and spheres quite appropriate for laser work and samples have been available for at least ten years. None the less such refinement of technique would not be necessary if these crystals were intended for ornament.

Hydrothermal Rubies

Hydrothermal growth has already been mentioned, but the need for a sealed pressure vessel would be seen as a hindrance to crystal growth on a large scale. In addition, speed of growth (on seeds) is critical if veil-like inclusions are to be avoided and the vessel needs to be free from iron. In practice this means that a precious metal lining has to be used.

Gübelin, in a report in 1961, described hydrothermally grown ruby, noting that the process was similar to that used routinely for the growth of quartz. Compared with quartz, however, corundum growth by this method poses some problems since aluminium oxide can exist in such modifications as gibbsite, böhmite and diaspore as well as corundum. The nutrient used in some experiments was poorly crystallized gibbsite or corundum. This was placed in the bottom of the autoclave, which was lined with silver and filled with sodium carbonate. Seed crystals of natural or synthetic ruby were placed in the top of the vessel, suspended from a silver frame. The autoclave was heated from below to 400°C, with growth on the seed plates taking place by convection. One or two months' growth might be needed for the completion of reasonably sized crystals for ornamental use.

Hydrothermal rubies grown by Chatham gave a normal ruby RI of

1.76–1.77 with a birefringence of 0.008. It is possible that if the seeds on which growth took place had been made from synthetic ruby the rather low degree of phosphorescence seen after X-ray exposure would have been more prominent. The seed contained natural inclusions but could be seen to stop at the junction between seed and overgrowth. Profuse minute gas bubbles could be seen in the overgrowth. Crystals intended for the lapidary were issued by Chatham in 1966. The specimens reported were a deep purple-red with a noticeable seed present. Transparency to SWUV showed the seed prominently: stones were more transparent to SWUV than natural rubies.

Corundum Ribbon

While it has been found possible to grow thin flat corundum crystals in ribbon form I have not yet seen reports that they have been grown in colours suitable for ornament. Such thin layers could be used as the upper surface of a doublet or as the background of a watch dial for which synthetic ornamental corundum cut from Verneuil boules has been used for many years. This method of growing corundum ribbon is known as edge-defined film-fed growth but the finer details are outside the scope of this present book.

Linde Star Rubies

Verneuil-grown star stones were produced by the Linde Division of Union Carbide Corporation in 1947. The asterism is caused by very small crystals of aluminium titanate. These form sheets which are aligned at 60° to one another. The addition of rutile to the feed powder achieves the desired effect. The best stars (according to the firm's publicity material) are formed when the proportion of rutile to the total feed powder is from 0.1 to 0.3 per cent and the boules are kept at 1,100–1,500°C for several hours to allow the needles to crystallize out.

Ensuring that the centre of the star is placed at the centre of the cabochon is not easy and the star needs to occupy the whole of the top of the stone. Fluctuations in the oxygen supply and consequent regular fluctuation of temperature seems to be the key to success in this area.

Another Linde product produces the star effect on the surface of the stones by polishing a ruby cabochon grown without rutile in the feed powder. The stones are then subjected to surface diffusion of rutile which forms a layer of needle-like crystals. The stones are then given their final polish. These products show a greater transparency than other synthetic star stones. The year 1952 seems to mark a change from the production of high-transparency star stones to those of lower transparency. Curved striae were more pronounced in the earlier products.

Linde stars were reported in purple, green, pink, yellow and brown as well as in ruby and blue sapphire colours. They were graded from A to C according to their depth of colour. One of the dichroic directions of the ruby is more inclined to yellow than in natural stones.

Inclusion Inducement

We have seen that ruby can be quite successfully imitated by 'crackled quartz', and that the liquid-filled feathers or fingerprints of natural corundum can be mistaken for undigested particles of flux (or, of course, the other way round). From time to time attempts have been made to simulate natural inclusions in synthetic products but the practice seems never to have been widespread or to have taken place over many years. For this reason examples do not often surface. Both Chatham and Knischka reported to GIA that they had induced inclusions into their rubies at some time but that specimens were so treated in pursuit of researches into crystal growth and characterization. Lechleitner admitted that some of his rubies with induced fingerprint-type inclusions had been placed on the market.

The process of inclusion inducement resembles the production of crackled quartz. Specimens are heated and then rapidly cooled by quenching in a liquid or melt. This induces surface-reaching fractures. The rubies are then placed in a flux melt containing dissolved aluminium oxide. The flux enters the rubies via the fractures and cools once inside the crystal. The resulting fingerprints are composed of flux particles. It may be possible for Verneuil boules or boule sections to be grown in the normal way and then placed in the hot flux. This produces a flux layer which covers the Verneuil material. The flux material may show parallel growth lines and some crystal faces. On polishing the flux coating may be removed.

GIA has found that if some of the flux coating remains it may be possible to see small birefringent crystals in the boundary between the coating and the core. Their RI showed them to be corundum with a different crystallographic orientation from that of the host crystal. The Verneuil seed has been found to contain several examples of fingerprint-type inclusions closely resembling similar ones in natural corundum. Gas bubbles and curved growth layers show the flame-fusion origin. Traces of molybdenum and lead found by laboratories must have originated from a flux – gemmological tests cannot detect them.

In a paper published in 1994 GIA reported that flux-induced inclusions had been found in a sample of about sixty synthetic rubies that were passed as natural by trade sources. Under magnification and viewed by reflected light the stones showed a net-like pattern of a whitish to colourless substance that reached the surface. Under transmitted light these inclusions formed a continuous three-dimensional cellular structure like a honeycomb. Normal gemmological tests will not distinguish these rubies so their microscopic features need to be learnt.

It is clear that synthetic rubies should not give the experienced gemmologist serious trouble. Verneuil flame-fusion stones look suspiciously clear, and with careful attention to lighting should display curved growth lines and curved colour banding together with large, bold, rounded randomly scattered gas bubbles. Furthermore there will be no solid inclusions. Boules can be cut to resemble crystals of natural ruby. The finished stones are very cheap and quick to manufacture. Artificial star rubies are always made by the Verneuil process, and look too good to be true – if one is familiar with natural stars.

Much more expensive rubies are grown by the flux-melt process, and while they show no natural inclusions, they normally have at least some traces of the metallic-looking flux compounds. Flux can form structures closely resembling the fingerprints made by liquids in natural stones. Some growers have made rubies on seeds of natural or synthetic ruby or colourless corundum.

IMITATIONS OF RUBY

The attractiveness, scarcity and value of natural ruby, together with its hardness, have led to a variety of imitations being hopefully or unconsciously offered for the genuine stone. Glass is naturally the most likely artificial product to be offered but in my experience one of the most successful is quartz. Rock crystal is of course colourless but if it is heated and then rapidly quenched in a cold liquid, surface-reaching cracks will develop. If the liquid used for the quenching contains an appropriate red dye, this will colour the whole stone red. The resulting colour is quite close to that of ruby; the rock crystal is faceted before treatment so that the lapidary can eliminate any give away natural inclusions peculiar to quartz. An emerald imitation is achieved with the same material in the same way. Physical and optical properties will of course be those of quartz rather than corundum.

The gemmologist will gradually acquire a familiarity with the various forms of glass and their properties, which we shall look at more closely later in the book. For the present we should remember that glass will not share any of ruby's properties. It will be much less hard and durable, so that if it has been faceted it is very likely that the girdle and facet edges will show small and very characteristic conchoidal (shell-like) fractures. Moreover, as glass is not crystalline, it can show no dichroism. Its refractive index will vary widely from specimen to specimen and stones which can be tested on the refractometer will show a single shadow-edge anywhere between 1.50 and 1.70 compared to the two edges shown by ruby at 1.76–1.77. Glass will not show a chromium absorption spectrum.

We have already discussed the quite effective ruby imitation provided by crackled quartz but it can also be imitated by composites in which a slice of almandine (dark red) garnet is fused to a glass base (which is in fact the bulk of the stone as the junction between the two portions is not at the girdle). Gemmological testing will quite easily detect this kind of composite. The spectroscope will give the (iron) almandine spectrum rather than the chromium one and the ruby refractive index will be replaced by a negative reading.

The Verneuil-grown red spinel makes a good imitation of the darker red rubies but owing to production difficulties (boules tend to fragment) specimens are among the rarer synthetic gemstones. Gemmological tests easily distinguish red spinel from ruby. Flux-grown red spinel from Russia is rather too dark to be confused with ruby.

We have already noted the colour enhancement of some darker red Thai or perhaps Thai-type rubies whose iron content inclines the 'pure' red to a darker brownish-red. Heating drives off some of the iron to leave a more Burma-like red. Though the heating process may well have distorted the inclusions (they are sometimes referred to as 'exploded') their nature should tell the gemmologist whether the specimen originated in Myanmar or Thailand. For certification purposes this is important information.

SYNTHETIC SAPPHIRE

From the beginning we should remember that sapphires are not always blue. They can be any colour but red (these are rubies) and colourless specimens are occasionally mined but more frequently grown by the Verneuil flame-fusion process. More effort is put into ruby growth since ruby commands higher prices than any of the sapphire colour varieties. Sapphires in nature are much more plentiful than rubies, although fine sapphires (of attractive colour, not necessarily blue) are also rare and desirable.

While chromium causes the colour of ruby, it plays only a minor part in sapphire coloration. Blue sapphires obtain their colour from an interplay between iron and titanium, both being necessary for the colour to form. Iron is the cause of colour in green sapphire and in some yellow sapphires. Chromium may play a part in the coloration of some purple/mauve sapphires and in some pink-orange specimens.

The importance of blue sapphire can be seen by a careful reading of jewellery auction sale catalogues, where important specimens have their country of origin stated – Myanmar, Sri Lanka and Kashmir are the only ones to be used so far, though fine Madagascar and African stones may reach this category in time. It is interesting to note that 'Burma' is the only country of origin given for ruby.

Despite the importance of blue and 'fancy' (any colour but blue) sapphire, crystal growers have not found much commercial success in using any technique other than flame-fusion growth. There are a few specimens of flux-grown and even hydrothermal sapphires reported in the literature and I have seen some of them but in general growers do not seem to have been able to control inclusions so that blue, pink and orange varieties (the only ones with which any real effort has been made) all carry profuse flux veiling of the type already described in our account of ruby. On the other hand, the Verneuil technique is cheap and apart from the curved growth lines and colour zoning, gives a clean (inclusion-free) stone.

Verneuil sapphire. Gas bubbles and irregular 'hose' gas inclusions

Verneuil sapphire. Curved growth zones, emphasized by numerous tiny gas bubbles

The oval sapphire is 'virgin'. Those on either side are diffusion treated

Verneuil sapphire. The Plato-Sandmeier effect between crossed polars

Some hydrothermal rubies grown on seeds have been found to show profuse gas bubbles on the seed coating. Sometimes the seed can be seen as a whitish inclusion beneath the red overgrowth. One Gilson experimental crystal I have seen shows this effect very well.

As far as identification goes the Verneuil-grown blue sapphires do not usually show the absorption bands at 470, 460 and 450 nm which are characteristic of most natural sapphires. Some blue stones have shown faint traces of the 450 nm absorption band but it would not be expected in pale blue or yellow specimens. These bands are due to iron which can of course be excluded from the melt. Most of the fancy-coloured sapphires hardly show any curved growth lines, and even though they are easier to see while the specimen is immersed in a liquid with a matching or close refractive index, the lines may still be hard to make out. They will sometimes reveal themselves if photographed. Today it is not possible outside a laboratory to obtain liquids with high refractive index though di-iodomethane serves quite well.

Natural blue sapphire will provide star stones when appropriately cut, provided that sufficient titanium dioxide (the mineral rutile) is

Early hydrothermal ruby. Natural seed ruby with polysynthetic twin lamellae surrounded by synthetic coat (mantle)

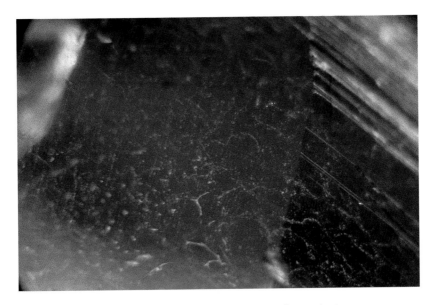

Early hydrothermal ruby. Telltale features in the synthetic coat, formed by gas bubbles

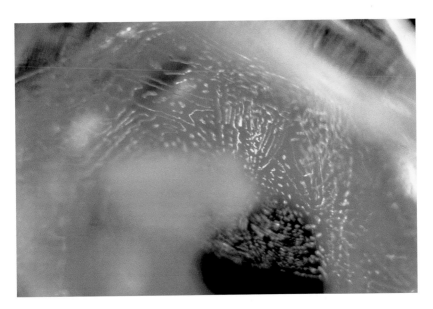

Early hydrothermal ruby. Deceiving liquid 'feather' (fingerprint inclusion)

present. This is necessary to provide the blue colour, so it is invariably present, but is not necessarily found in the other, fancy sapphire colours. This is why yellow or green star corundum is not known. Star blue sapphire, like star ruby, is obtainable only from Verneuil-type crystals or by diffusion. The finished stones, again like the rubies, show too-perfect stars with rays meeting in the centre of the cabochon and usually with flat polished backs. Examination under the microscope will show the customary large gas bubbles, though in darker specimens a strong light will be needed and the edges of the stone will have to be examined rather than the centre.

Blue sapphire grown hydrothermally in Russia by Tairus is reported to have used nickel as a dopant instead of iron and titanium. The colour is said to be sky-blue; different colours have been grown with the addition of nickel, or other elements and varying the concentration. Greenish-blue sapphires are said to show a red fluorescence under UV and to contain crystalline copper inclusions. Nickel-doped greenish-blue sapphires show an absorption spectrum with three intense bands at 970, 599 and 377 nm. with two weaker bands at 556 and 435 nm.

Blue sapphire crystal groups grown by Chatham were made up of

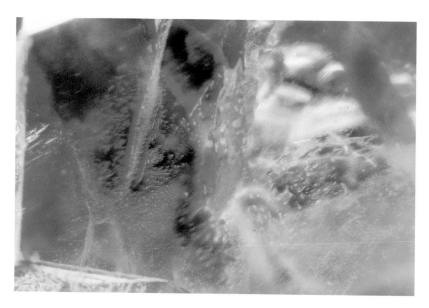

Chatham sapphire. Characteristic scenery consisting of tousled fingerprint inclusion

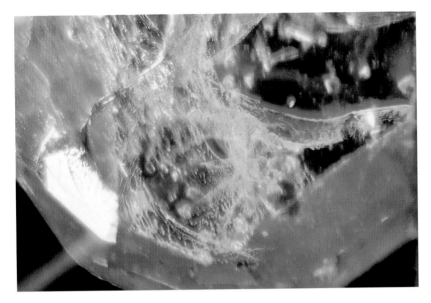

Chatham sapphire. Dense accumulations of fine liquid films and feathers

Chatham sapphire. Concentration of typical platinum inclusions

randomly-arranged tabular crystals. Colour was concentrated in the crystal tips, the remainder of the crystals being colourless. The blue ranged from medium dark to very pale and showed strong angular banding. Dark angular platelets of crucible material could be seen and twisted flux veils were prominent. Under LWUV the sapphires showed an overall yellow with dark blue and greenish patches. Under SWUV the crystals were an overall blue, although parts of the surface and the junction areas of some of the crystals were yellow.

Yellow Sapphire

While the reasons for growing ruby and blue sapphire are obvious, the other colours of sapphire are less sought after. Of those yellow is probably the most popular, although not every country admires this colour. Fine yellow sapphires can command high prices so that the gemmologist, when faced with one, has not only to establish the unknown's species (which can be done by normal gemmological tests) but also to establish whether the stone is natural or artificial. Even then, the nature of the colour still needs to be determined. There are several ways in which natural or synthetic corundum can take a yellow colour: this would not be a problem if it were not for the fact that some yellow sapphires may fade in strong sunlight. There is no way in which any yellow sapphire can be tested in advance to see whether or not it may fade, apart from a fade test.

Natural yellow sapphires owe their colour to iron, although in many Sri Lanka stones, which give an apricot fluorescence under LWUV, the iron content would have to be low. In synthetic yellow sapphires the colour may be caused by doping with nickel or with nickel and iron to give a greenish-yellow. The imitation of padparadschah, which inclines to an orange-pink, may be doped with nickel and chromium with some iron. None of these colouring elements affects measurable properties, so the gemmologist has to rely on the spectroscope (to detect iron and chromium, though this can be difficult) and on the microscope for evidence of Verneuil-type growth. Even with a microscope detection may be difficult, because in yellow Verneuil sapphires the curved growth lines are almost impossible to see. The best technique is to photograph the stone, as then the lines may be visible. Experienced gemmologists will also suspect the absence of natural mineral inclusions.

A synthetic hydrothermal yellow sapphire was grown in China on a seed of colourless synthetic corundum, the colour arising from the impurity elements nickel, cobalt and chromium. Increasing the nickel content gave a darker green, inclining to brown. Pink, near padparadschah colours were obtained presumably by the edition of chromium. Two-phase inclusions, planes of gas bubbles and acicular crystals have been reported; in addition, wedge-shaped intersecting growth banding and colourless seed plates are also known. The report appeared in the Chinese *Guilin Journal of Gems and Gemmology*, 31(1), 2001.

Pink Sapphire

It is often said that whether or not a red to pink corundum specimen is ruby or pink sapphire depends upon whether or not you are buying or selling it. None the less pink sapphire as a distinct colour variety certainly exists and at the time of writing it is noticeable in top-rank London jewellers.

It is not surprising that attempts should be made to grow saleable specimens which might be thought to appear more realistic than the Verneuil product. In 1995 Union Carbide Corporation published details of a pink sapphire ('pink Ti-sapphire' – the Ti is titanium) grown by pulling. The pink was one of several colours grown at that time. The pink colours inclined either to orange or to purple and stones showed a rather faint fluorescence under LWUV. Under SWUV stones showed a weak to moderate bluish colour with some chalkiness, an effect not seen with natural or synthetic chromium-doped corundum. Almost all synthetic pink sapphires show a chromium absorption spectrum, but the Ti-sapphire shows only faint indeterminate bands. Titanium alone has not so far been reported as a colouring agent in corundum.

Under magnification Union Carbide pink sapphires showed pinpoint inclusions and very small gas bubbles in high relief – these features are characteristic of crystal pulling. On immersion some of the sapphires showed very faint colour banding and in some cases the effect is best seen using polarized light. No curved colour banding was ever reported.

If fluorescence effects were the first to be tested the Ti-sapphires should not present too many problems but gemmologists usually turn first to a refractometer – a microscope would be preferable. Any suspected red to pink corundum should be checked with a spectroscope.

Other Colours

Green flame-fusion sapphire has produced some faceted stones after the boules have grown in a dark blue cobalt colour which after crystallization turns to green, although one end of each boule retains the blue colour. It is likely that cobalt or a combination of cobalt and vanadium is the cause of the colour.

We should remember that many 'alexandrites' are in fact a vanadium-doped synthetic corundum, which shows a colour and a colour-change which turns from a slaty blue to a purple when examined by daylight and incandescent light respectively. This is a Verneuil product which shows the curved growth banding especially well and also a diagnostic absorption band at 475 nm. Though rarely mentioned in the textbooks, these stones look much more like amethyst than alexandrite.

ENHANCEMENT OF RUBY AND SAPPHIRE

Clarity Enhancement

Rubies and blue sapphires are sometimes improved in clarity by having surface-reaching fractures filled. As in diamond and emerald the flash of a completely foreign colour, perhaps green or purple, should alert the tester to the possibility of filling. Neither ruby nor blue sapphire seems to have been oiled to the same extent as emerald where oiling is routine today but I would not rule out the odd example turning up.

Beryllium-doped Sapphires

The journal *Gemmology Queensland* for October 2002 has reported on a sapphire treatment which probably originated in Thailand. After treatment, the stones have assumed an orange-pink padparadschah colour (this has always been a good selling name even though no two people seem to agree on the exact colour). On examination they have shown the presence of beryllium, which has been diffused into them. Experiments appear to have shown that when used to treat stones in bulk (this point is important) beryllium does affect the colour. John Emmett and Troy Douthit, whose knowledge of and experience with the colour enhancement of corundum is unrivalled, have supported the bulk diffusion of beryllium colour enhancement theory against at least one alternative, the internal migration of electrons during treatment (EM).

The journal summarizes the evidence supporting bulk beryllium diffusion via HP/HT treatment as follows. Diffusion creates a yellow-orange colour in corundum. Previously colourless tones have changed to yellow to gold, pink stones have changed to a pinkish-orange, red stones to an orange-red and green ones to yellow to gold.

When the beryllium has incompletely diffused through the stone it is possible to see a yellow to orange rim at and below the surface of the stone. This is best observed when the stone is immersed in di-iodomethane and examined with diffuse light-field illumination. The effect is seen best in the padparadschah pink-orange stones and also in some rubies.

When beryllium diffusion is complete and especially in yellow to golden-yellow specimens the treatments cannot be identified by standard gemmological testing. All stones seem to show the customary effects of HT treatment: melted surfaces showing recrystallized corundum are common.

A useful overview of the present position of orange treated sapphires can be found in *SSEF Facette International*, issue 10, January 2003. The orange sapphires first began to appear on the market in 2001, the colours ranging from strong orange to an orange-red. Lemon yellow sapphires have also been noted.

Treatment is carried out on low grade sapphires of different colours. Purplish, pink and white are believed to be the commonest colours. The sapphires are heated and originally this was done with topaz, zircon and chrysoberyl, which provided a source of beryllium. Later, chrysoberyl seems to have been the sole additive. Treatment takes place with borax and elements from the additives which are not part of corundum's chemical composition diffuse into the sapphires.

Beryllium diffusion causes a yellow rim within the stones to develop to a depth of several millimetres. Many stones showed surface-related zones of a pink colour while the centre showed pink. While the name padparadschah has naturally been used, it is inappropriate for colour-diffused stones of this type.

At a meeting in Tucson, Arizona, Chantaburi traders from Thailand explained that corundum from Madagascar and Tanzania was being used for treatment – they denied at that time that beryllium was being used. It was later shown, however, that chrysoberyl *was* used as described above rather than being a result of migration from the apparatus used for heating the treatment furnace.

The term to be used for the new material has given rise to some discussion. The names 'bulk-diffusion-treated sapphire', 'lattice-diffusion-treated-sapphire', 'heated sapphire' and 'heated padparadschah' have all been suggested, but none seem realistic at the time of writing.

An update on the same topic was published in the Winter 2002 issue of *Gems & Gemology*. GIA had re-examined some of their earlier work and the results previously published in *Gem News International* and in the *GIA Insider* (which can be assessed at www.gia.edu/wd_349.htm for the dates 28/1, 15/2, 3/5 and 1/11 2002). The additional investigations showed conclusively that the diffusion of beryllium into the crystal lattice of some sapphires was the cause of the change of colour and the concentration of the beryllium could be up to ten times greater in the rim of the sample than in the centre. Analysis of the original surface of the one treated orange sapphire showed a beryllium concentration of up to 99 per cent. When the beryllium had diffused throughout the stone no colour zoning could be seen.

Some experiments showed that colourless synthetic corundum of high purity could be turned a strong colour when about ten parts per million of beryllium was diffused into it.

The major journal paper and summary of what is so far known of the technique and results of beryllium diffusion of ruby and sapphire is by Emmett et al. in *Gems & Gemology*, Summer 2003. It includes an important and topical bibliography as well as a summary which can be quickly referred to under the headings of diagnostic evidence, highly indicative evidence, evidence that is not indicative and indications of no exposure to high heat.

The main points made by the paper include diagnostic evidence of

*Natural orange corundum is not common and large clear examples
should be treated with suspicion*

colour zoning conforming to the external faceted shape of the stone,
which is seen in all types of beryllium diffused corundum and can
penetrate to any depth. Immersion is needed.

The 'highly indicative' evidence includes highly altered zircon
inclusions. These may show alteration to white masses from which the
crystal form is absent (a sign that very high temperatures have been
used) or be present as small transparent angular or rounded grains.
Internal recrystallization is also a sign of exposure to high temperatures, discoid fractures around zircon crystals are common and a roiled
appearance can be caused by the destruction of boehmite crystals and
their replacement by recrystallized corundum in the channels they
formerly occupied.

Internal lattice diffusion around rutile crystals (some forming silk) is
common in heated blue sapphires, shown as roughly spherical blue
haloes round the crystals. This is now considered very rare in ruby and
yellow-orange sapphires, since heating takes places in an oxidizing
atmosphere which normally removes the blue colour. When seen, these
phenomena should be taken as a very likely result of exposure to high
temperatures. Some sapphires, those from Montana river gravels in

particular, which are nearly always heated, though not diffused, may show blue spotting but this should not be taken as a sign of beryllium diffusion.

Synthetic overgrowth can frequently be seen in high-temperature treated rubies and sapphires, occurring as minute platelets which may not always be polished away. These seem to be most commonly seen on yellow to orange-red stones and give a clue to high temperatures.

Synthetic crystalline material grown during beryllium diffusion treatment is usually randomly orientated with a hexagonal shape. The authors noted increased synthetic growth on the surfaces of some Mong-Hsu rubies.

Inconclusive evidence included highly saturated colours which, although obtainable from beryllium diffusion can be repeated by some natural material. Healed fractures may arise during heat treatment without diffusion. Dissolution of the surface is very common in beryllium diffused stones but again they are commonly seen in heated stones in which diffusion has not taken place.

Stones which have not been exposed to very high temperatures may show internal voids containing a bubble of carbon dioxide and water. This feature is particularly common in Sri Lanka corundum. Such inclusions cannot survive very high temperatures. Undamaged zircon crystal inclusions would suggest no heat treatment, nor can rutile needles retain their form during exposure to very high temperatures.

A sapphire cut as a faceted stone, weighing nearly 120 ct and with an orange-yellow colour, was examined by GIA and the report published in the Fall 2002 issue of *Gems & Gemology*. A new technique known as lattice diffusion was found to have been used on this stone. Properties were consistent with sapphire but the stone gave a weak orange fluorescence under LWUV and a moderately chalky but very weak patchy yellow under SWUV. When immersed in di-iodomethane the stone showed a near colourless core with an orange-yellow zone next to it, which followed the outline of the facets.

Though beryllium, found in similar yellow-orange layers in bulk diffusion-treated sapphires was not thought to be a cause of colour on its own, it is now believed that it may be so when diffused into the stones in an oxidizing atmosphere. There were no Plato lines in this stone and high gallium and titanium indicated natural origin.

Bulk diffusion of yellow colour into sapphire requires high temperatures, which can create specific changes in the material. Some
corundum may partially dissolve in the crucible when it is in contact
with the flux used and synthetic corundum can then grow on the
surface of treated stones as they cool. Redeposition of corundum is
now becoming familiar on the surfaces of preforms and on treated
rough, and often shows as groups of very small flat hexagonal
platelets. Polishing will of course remove these growths but Scarratt
has noted on the AGTA Gemmological Testing Centre web site
(www.agta.org) that some corundum of this kind was left on polished
specimens. The illustration of some of these crystals, in *Gems &
Gemology*, Fall 2002, shows (in polarized light) randomly oriented
crystal groups. The random orientation means that individual crystals
will show different directions of extinction from one another as well as
from the host. If the host is placed in the dark position, some of the
crystals will show light.

A blue sapphire examined by GIA after its identity as sapphire was
established was found to show a shallow and patchy blue coloration
with outlining of facet junctions. It was also noted that in some orientations the stone showed curved colour banding extending across its
width. The banding was a pale blue and was probably the colour of the
original material.

This coloration was the product of bulk diffusion (the original
colour diffusion process, patents for which date back to the mid-1970s
as a description in *Gems & Gemology*, Fall 2002, reminds us).

The same issue of *Gems & Gemology* describes a 0.95 ct light yellow
pear-shaped heat-treated sapphire which was deemed to be natural by
standard tests. Inside the stone were small light-scattering particles in
angular formation and the facets were covered with solidified droplets
of a transparent substance about 6 in hardness. This substance seemed
to be amorphous.

The table facet showed clear signs of high-temperature treatment
with flow lines and ridges easy to see with shadowed fibre-optic light.
The coating was found to be a glass. As the stone had not been repolished after treatment the tell-tale signs were much easier to see. It is
worth checking the surface of doubtful specimens to see whether
remains of the surface layer are visible on unpolished areas.

A note in *Gemmologie* 51(1) 2003, throws further light on beryllium-treated sapphires. John Emmett stated that a completely colourless stone with a diameter of around 7 mm could be produced in less than 100 hours of heating. The German Gemmological Association tested a sample of faceted stones covering a weight range of 0.09–0.20 ct, with colours ranging from pink through pinkish-orange to an intense orange-red.

When the stones were immersed a pink core with an orange rim could be seen, characteristic of the beryllium-treated sapphires already published.

Some specimens, however, were either completely coloured or showed a colour zoning parallel to the growth zoning rather than the outer shape of the stone.

Identification of the treatment, which would otherwise pose problems, was possible in many instances because spherical blue zones surrounded inclusions of rutile. These arise from the internal migration of iron and titanium and are typical of beryllium-treated sapphires from Songea, Tanzania.

Diffusion Ruby

The Fall 2002 issue of *Gems & Gemology* described a ruby, said to emanate from Bangkok, which had been on the market for some time. The producers advertised the specimens as diffusion-treated corundum, with a shallow colour. The paper showed how this material was in fact a synthetic ruby overgrowth on natural colourless to near-colourless corundum. The overgrowth layer, after recutting, turned out to be from 0.1 to 0.3 mm thick.

To the eye the faceted stones appeared a saturated red colour when face up. A white background for this test showed that the colour distribution was uneven or patchy in some specimens. For some specimens gemmological features (RI, DR, SG) were in the corundum range (RI from the table in the range 1.760–1.761, 1.768–1.770 for the ordinary and extraordinary rays respectively), although some RIs were harder to read than usual. On the other hand, the majority of the specimens tested showed RI over the limits of the refractometer liquid (cited in the paper as 1.81). This unusual feature was seen in all the pavilion facets tested. The colourless core was found to have an RI of 1.760–1.768.

The stones all showed a characteristic ruby absorption spectrum, elements ranging from weak to very strong. All specimens were inert to faint red when under LWUV. One stone gave faint orange fluorescent zones in the interior only. No stone responded to SWUV.

Under the microscope some of the specimens showed areas of the original surface of the outer layer which had not been removed when the stone was repolished. These areas showed undulating though smooth contours and stepping, with small planar surfaces which looked as though multiple crystal faces were developing. Cavities or indentations were also seen, concentrated in some areas and sometimes down to the full depth of the shallow outer layer, without ever reaching the interior of the stone.

The patchy colour zoning between adjacent facets was easily seen under magnification, especially when a diffused light was used. In all of the samples the red coloration stopped abruptly at a boundary plane which could be seen just below the surface. The bleeding of colour between outer and inner regions, seen in some diffusion-treated rubies and blue sapphires, was not visible in any of the samples. One stone showed a yellow to orange colour zone which was restricted to one portion of the interior.

The microscope is the best tool for the investigation of diffusion-treated stones and in this case the unusually high refractive index readings would also arouse suspicion.

Enhancement of Rubies and Sapphires Using a Flux

In early 2004 Ted Themelis published a short but informative guide to the enhancement of rubies and sapphires using fluxes, *Flux-enhanced Rubies and Sapphires*.

While heat treatment has been used for many years in the removal of unsightly inclusions to ensure greater clarity the processes used have not been completely successful, although they have been accepted by the gem trade. The filling of surface factures also helped to produce greater clarity. Themelis makes an interesting point, which I have not previously come across: that many Western dealers regard rubies and sapphires that have been heat treated with artificially induced fillings as 'semi-synthetic'. This naturally poses problems for certifying bodies, and gem-testing laboratories in particular. Themelis's book addresses this problem in so far as he describes the flux additives and other chemicals used in the enhancement process: he also looks at the nature of filling and describes some details of the heating used.

The stones used for the experiments outlined in the book were in the main chosen from Mong-Hsu and Mogok in Myanmar, Morogoro and Songea, Tanzania, the John Saul mine in Kenya, Luc-Yen and Quy-Chau in Vietnam and sites in Madagascar and Sri Lanka.

Themelis opens with some questions and answers to help the reader to get into the author's mind so that both are on the same track, with as few misunderstandings as possible.

Flux treatment is described as a thermo-chemical process using fluxes and other chemicals. It is carried out to enhance clarity, particularly in view of the scarcity of high-quality rubies and sapphires. The products of the treatment are marketable. He says that nearly all the rubies on the market today are heat-treated. Some pink sapphires also owe their appearance to heat treatment. Any ruby or sapphire with surface-reaching fissures can be treated and the methods used vary considerably. The commonest fluxes used are borax, often combined with silica. The finished stones are stable to light and to normal wear. Gemmological properties are unaffected by the treatment; even though the process can be detected the nature of the filling cannot be ascertained since some of the additive may be produced by the melting of natural minerals.

Disclosure of the treatment is mandatory and guidelines have been established by the United States Federal Trade Commission and CIBJO in Europe. However, he claims that many gem dealers and jewellers do not disclose known treatment for fear of losing sales. In the lower price ranges there is no difference in the asking price for untreated or treated material, but as the price rises differences become more apparent. Themelis believes that treated rubies and sapphires might cost half the price of untreated ones. Treated stones may not generally be seen as a good investment and treatment has been undertaken at least as far back as the early 1970s.

Fissures in the original ruby or sapphire arise from incongruent cooling of the melt and many different kinds of cavities, empty or containing minerals, are formed. The flux used for the filling should not set up an unfavourable reaction with the host; it should be chemically free from impurities and be water-soluble. It should have a melting point below 500°C and low viscosity. Themelis adds that it should also be cheap. He gives details of some of the fluxes.

Natural minerals can be used as additives: chrysoberyl has been extensively used in the beryllium treatment of corundum, as Themelis explains in another book, *Beryllium-treated Rubies and Sapphires*. Natural quartz is routinely used in the treatment of rubies from the John Saul mine in Kenya.

Specimens are cleaned before treatment, usually with strong acids, sometimes ultrasonically. After trimming and preforming the flux is applied to the stones, sometimes with a brush, and they are placed in a stainless steel container, passing from there to the crucible which, in Thailand (the centre of these operations) is, for rubies, a high-temperature electrical resistance furnace. Sapphires are generally treated with or without fluxes in ultra-high-temperature gas combustion furnaces. Themelis gives a general note, with no details, on heating and cooling, the latter involving no more than turning off the switch.

Most heaters remove the crucible from the furnace below 1,500°C and the stones from the crucible at about 1,200°C. Thermal shock may, it is believed, be avoided or alleviated when a borax-based additive is used as it acts as an insulator. The removal of the stones from the crucible may pose problems, as specimens could adhere to one another. Themelis describes several of the cleaning processes, some of them involving acids. Complete removal of the fillings from surface-reaching fractures has not been possible.

The response to this treatment by gem-testing laboratories varies, some ruling that if residues of the filler cannot be seen under 10x magnification and there is no reference on the report either to the presence of the filler or to what kind it is, the stone will be said to have been heat-treated only.

Silica contamination may occur during treatment and net-like clouds of undigested flux can be found. Whitish undigested residues can sometimes be seen surrounding the opening of a fissure reaching the surface. This has been observed in treated rubies from Mong-Hsu, Myanmar.

Where rubies of fine colour are involved, silk (intersecting rutile needles) was being removed by heating in the late 1990s. Themelis reports on features seen in Mogok rubies after treatment: distorted mineral inclusions and networks of undigested residues of the flux used in treatment are accompanied by streams of liquid droplets forming fingerprint patterns. It is interesting to see that some of the drops reach quite large sizes. Dissolved rutile needles show as dots.

Themelis treated some Vietnam rubies, mostly waterworn crystals, and found that most of the stones turned a purplish-pink with considerable improvements in clarity, transparency and lustre. A sample of stones he purchased in Pakistan (though the actual place of origin was not given) were a dark red-brown with many surface-reaching fissures and deep cracks. They were heated with borax, some phosphates and other additives at 1,650°C for three hours. After heating they turned pinkish-red with a purplish tinge. Themelis comments: 'Junk in, junk out' in this case at least. Areas where the filled fissures reached the surface showed white.

Some Sri Lankan sapphires, originally an unspecified colour, turned a very dark blue after heating in a mixture containing borax, although some crystals did achieve the fine blue characteristic of much geuda material. Rubies from India described as 'sandpaper quality' turned red or pinkish-red, giving an improvement on the original colour; heating was with borax at a temperature of 1,650°C. Themelis states that in this instance at least a flux with borax appeared to be necessary for a red-pink colour to be produced.

Afghan rubies with numerous inclusions of calcite and mica, as well as other minerals, were heated with flux, turning the crystals red with no tinge of purple. Clarity and lustre were improved.

Rubies from the John Saul mine at Mangari Swamp, Kenya, are opaque and heavily included with mica; heating improved the colour and clarity very considerably. Improvements were also made in the rubies from N'Dofu in the same country. Many Kenyan rubies are mica-rich ('mica rubies'), although heating with flux was found to enhance the colour and remove a muddy appearance.

Themelis found that Madagascan orange-pink sapphire treated with flux showed a 'sweating' appearance from undigested droplets of flux.

The surfaces of treated corundum often show signs of recrystallization and these, with the clarity of the fissures, are used as parameters in grading systems. The stability of the filled areas does not appear to be affected by strong direct sunlight and no colour changes were observed after exposure to UV radiations. Heating and/or acid treatment during repairs may affect the fillings and recutting may expose the top of a fracture.

NOTES FROM THE LITERATURE

An ingenious imitation of star blue sapphire is reported in the Summer 2001 issue of *Gems & Gemology*, which was quite successful. The stone had a flat back and appeared bruted. Most natural star sapphires have domed or lumpy backs. The blue colour was concentrated in a thin layer on the base, about 1 mm thick and containing profuse air bubbles which gave a grainy appearance. The upper part of the cabochon was made from clear colourless material in which were contained three sets of minute colourless needles arranged at 120°. These were the cause of the asterism.

The stone showed no natural inclusions, although there were some conchoidal fractures – not common in corundum. There was no trace of the 450 nm absorption band or of any other bands in the visible. However, between crossed polars the 'bull's eye' effect was easy to see and the stone was identified as quartz, the identification being confirmed by EDXRF examination and also by Raman spectroscopy. The colour layer turned out to be a blue enamel.

In *Journal of Gemmology*, 28 (2002), Duroc-Danner described a Verneuil-grown blue sapphire showing an iron absorption spectrum. Received wisdom on the presence or absence of iron absorption bands in blue sapphire varies with locality (natural stones) or type (method of growth and dopant). The cause of colour in blue sapphire is now held to be due to a charge transfer between iron and titanium.

The specimen discussed in the paper was a 7.02 ct oval brilliant-cut sapphire with properties in the normal range for this material. The absorption spectrum as viewed with the direct-vision spectroscope (in the visible light portion of the spectrum) included an absorption band, ascribed to iron, at 450 nm, seen in the direction of the optic axis, which in this specimen was parallel to the table facet. Also seen, using the UV-visible spectrophotometer, were two absorption bands, one at about 703.5 nm (extraordinary ray) and another at near 585 nm (ordinary ray): some literature gives 565 and 700 nm for these bands. Other absorption bands were detected at 450 nm in the ordinary ray and at

388, 377 and 328 nm (these three bands cannot be seen with the hand spectroscope).

The sapphire fluoresced a strong chalky-green under SWUV and was inert to LWUV. Polishing marks (fire marks) could be seen under magnification near to facet junctions on the crown and pavilion. These marks are like small wavelets seen when the incoming tide peters out on a flat sandy beach. Blue pinpoint-like inclusions, some appearing to form fingerprint-like shapes, could be seen under the table. Two small healed fractures filled with a glassy material could be seen close to the surface in the girdle area.

Large, well-shaped gas bubbles, universally taken to be diagnostic for flame-fusion products, were also observed. When immersed in di-iodomethane and examined between crossed polars in the optic axis direction two sets of straight twinning lamellae could be seen, intersecting with each other at 120° and 60°, the Plato effect which is characteristic of Verneuil corundum.

Broad, uneven growth bands from colourless to deep blue could be seen parallel to the optic axis direction when the stone is immersed.

With the thought that the stone might have been diffusion treated the colour concentrations along facet junctions were checked while the specimen was immersed but no concentrations were noted. The blue pinpoint inclusions were thought to originate from the powders used as feed material; they may not have mixed evenly.

It is probable that the high temperature of the Verneuil process began to melt the (probably coarse) powders with their trace element contents and that some form of accidental diffusion took place when these elements interacted with one another before fusing. This would also explain the presence of the otherwise unlikely iron absorption spectrum, which is rarely seen in Verneuil blue sapphires. This absence is due to the burn-off of most of the iron and titanium in the Verneuil furnace.

A ruby tested and found to be a flame-fusion Verneuil synthetic stone contained fingerprint-type inclusions characteristic of natural rubies but not of this type of artificial product. In 2003 Duroc-Danner described an oval-shaped faceted ruby of 2.60 ct.

The stone was taken from a parcel of rubies whose origins were to be determined. It gave the normal RI and SG for ruby as well as the expected absorption spectrum and fluorescence. When immersed in di-iodomethane and viewed between crossed polars the stone did not show the Plato effect of two sets of lines intersecting at 60°.

Under magnification many distinct curved growth lines were observed immediately through the table facet and small groups of clouds consisting of pinpoint-like gas bubbles trapped between growth lines and confined between two zones on one side of the crown near to the edge of the girdle.

There were also, and significantly, a number of veil-like fingerprints, wispy and twisted, resembling those seen in the Mong-Hsu rubies from Myanmar. These were quite easy to see through the crown. Several of them broke the stone's surface.

Some of the fingerprint-like inclusions were filled with a glassy dark red substance (the author speculates on the possibility of their being flux inclusions reflecting red from the very strong saturated red of the stone). A flat fingerprint was seen on the surface and showed net-like patterns like those seen in flux-grown corundum.

A small flat surface fingerprint has a folded formation very like structures seen in Burma sapphires. There were also some parallel straight polishing lines which met at an angle between two adjacent facets.

The practice of inducing fingerprint-like inclusions into Verneuil synthetics has been reported from time to time. One way is to heat and quench the specimen, and the protrusion from the surface of some of the fingerprints strongly suggested that this had happened.

The net-like patterns shown by the flat fingerprint at the surface and the folded effect noted on the small flat structure at the surface were very deceptive and could be mistaken for similar features in natural ruby. The induced fingerprints helped to hide the curved growth lines which were easy to see on this occasion (not always the case) and the polishing lines meeting at an angle on two adjacent facets could have been put there deliberately.

Absence of the Plato lines is not uncommon in Verneuil rubies, especially if they have been heat treated. The fingerprints were induced, it was concluded, by quench cracking.

It is significant that this specimen was taken from a parcel of rubies which had already been cut. This suggests that the other stones in the parcel may have been of quite respectable quality. It is unlikely that the practice is seriously widespread but gemmologists should never console themselves with the thought that no one could make a successful business from a particular kind of treatment.

A report in *Materials Chemistry and Physics*, 66, 2000, reminded readers that over 90 per cent of the world's synthetic corundum is produced by the Verneuil flame-fusion process. Boules have always tended to crack along their length as the crystal cools. The authors of the paper examined crosswise and lengthwise slices around 1 mm in thickness by optical microscopy, energy-dispersive spectrometry and X-ray transmission topography. They found that only the bottom portion of the boules were single crystals. From the middle upward there was a good deal of stress deformation, easily seen as interference colours between crossed polars; this deformation leads to a macromosaic structure. The authors believed that the loss of crystallinity and build-up of stress led to the cracking. Areas of the growing crystal nearest to the flame underwent a type of annealing which improved their crystal quality.

A short note in the Summer 2001 issue of *Gems & Gemology* described a thermal enhancement scale for corundum devised by the Gübelin Gem Laboratory, which was intended to characterize and classify the whole range of heat-induced features in corundum. The scale will be used as an adjunct to whatever type of disclosure is used. Master stones are used for comparison.

There are six points on the scale on which TE1 is used to describe minute traces of residue inside partially heated fractures. TE6 would indicate a treated stone in which large glass-filled cavities were prominent as well as healed fractures with drops of residue. The system goes some way towards explaining what the enhancement has achieved rather than stating the amount of enhancement.

Sapphire from the Subera area of the Central Queensland gemfields, Australia, is not of the highest quality and heat treatment of the material in bulk is described by Maxwell in *Australian Gemmologist*, 21, 2002. Much of the sapphire is parti-coloured but the author notes that not all treated material increases in value and some in fact decreases; this experience is worth reading about as it goes against a common assumption that treatment always enhances a gemstone's appearance, otherwise it would not be attempted. The proportion of successful treatments to unsuccessful ones (from the commercial standpoint) is about 50:50.

The project was developed deliberately as a low-cost operation, as labour is expensive in Australia. It was hoped that stones would not have to be sent for heating to Thailand, as they were originally.

The process was devised by the American firm of Crystal Chemistry (formerly Crystal Research). Full details are found in the paper, but a summary may be useful. It was not found to be advisable to treat good-coloured blue sapphire crystals, whether dark or light blue. Good-quality green sapphires (light or dark) should not be heated beyond the first stage of the process (acid washing). (After acid washing, the full process involves the sorting of the gem material into colours and grades and the application of one or more heat treatments.) Second-grade light greens should be treated no further than the first stage. Yellow-green sapphires should not go beyond the first stage and yellow stones should not be treated at all. The above relates only to gem material.

Non-gem corundum heat treatment was also tried, and colours were blue-black-grey, green, dark brown, mid-brown and light brown. Stones were put through one, sometimes two or three, heating processes. The significant results in general were that blue-black-grey, green and dark brown sapphires produced some blue and green specimens, while dark brown material produced some blue and green.

Probably the most interesting conclusion of this experiment was the unsuspected potential of the dark brown stones, which could not have been guessed at prior to heating. A question asked by Maxwell was 'As most Australian heat treaters only treat medium to good gem quality sapphires ... why do Thai heat treaters seek corundum so desperately?'

A good general overview of the hydrothermal corundum manufactured by the Tairus joint venture in Russia can be found in *Gemmologie*, 52(1) 2003. The authors found that red, blue, violet-blue, greenish-blue and orange corundum had RI maximum birefringence and SG corresponding with natural corundum. The measured values were 1.759–1.769 and 1.768–1.776 for the extraordinary and ordinary rays respectively with a birefringence of 0.007–0.010 and SG 3.98–4.02.

Ultra-violet/visible spectra showed both nickel and chromium as colouring elements. Chemical analyses showed distinct chromium and nickel contents in addition to iron. The infra-red spectrum showed that oxygen-hydrogen groups were present as well as various carbon-oxygen groups which distinguish the Tairus products from other synthetic ones and from the natural corundum. Inhomogeneous growth phenomena are diagnostic for hydrothermally grown synthetics.

The hydrothermal ruby gives a chromium absorption spectrum and characteristic hydrothermal growth lines. A chromium absorption spectrum is also shown by some at least of the blue hydrothermal sapphires; these stones show a reddish colour when illuminated from a strong light source.

A ruby with a filling large enough for its refractive index to be measured was described in the Spring 2003 issue of *Gems & Gemology*. The purplish-red stone weighed 2.50 ct and was established as natural. With a 10x lens it was not difficult to see that the stone had been heat treated, as a series of partially healed fractures were detected as well as altered inclusions. The healed fractures look very like the fingerprint inclusions found in natural corundum.

During the test some opaque brown particles with a rounded shape were noticed. These were established as crystal fragments which had exploded under the heat treatment, the melted fragments assuming the rounded form on cooling.

In the area of the culet and deep inside the specimen large gas bubbles could be seen close to the surface. When the stone was examined with overhead lighting very fine separations could be seen in three of the pavilion facets next to the culet; these particular facets showed a different lustre from their neighbours. The RI of this area was found to be 1.51, indicating a type of glass.

The owner of the stone was told that it had been both heated and filled with a non-ruby material. If the stone was recut to remove the area including the cavity the loss of weight would be significant with a reduction to less that 2 ct.

The note stated that although such fillings are less commonly encountered than they would have been ten years ago it is always necessary to examine the surface of rubies and it also makes the point that if the stone had been examined through the table the presence of the gas bubbles would have given the impression that it was a Verneuil-type synthetic.

Two stones examined by the GIA East Coast Laboratory had crowns of natural green sapphire with synthetic sapphire pavilions appearing greenish-blue in daylight-equivalent lighting and reddish-purple in incandescent lighting (*Gems & Gemology*, Summer 2003). It was easy to see the two different portions when the stones were examined from the side and the purple base overwhelmed the green of the crown. If the join were concealed by the setting the deception could be troublesome. Curved colour banding could be seen in the pavilions and gas bubbles gave a conclusive result to the test. The pavilions were found to show an absorption band at 474 nm and under LWUV showed a medium strong orange fluorescence with a medium orange under SWUV. The green sapphire crowns remained inert under both types of UV.

Some hydrothermal rubies grown on seeds have been found to show profuse gas bubbles on the seed coating. Sometimes the seed can be seen as a whitish inclusion beneath the red overgrowth. One Gilson experimental crystal I have seen shows this effect very well.

The Summer 2004 issue of *Gems & Gemology* describes a light yellowish-green synthetic sapphire of 8.85 ct with no apparent inclusions to the eye. No iron-related absorption bands were detected with the desk-model spectroscope. EDXRF spectroscopy showed the colour to be due to cobalt. There has been no report of a natural sapphire owing its colour to cobalt.

CHAPTER SEVEN

Emerald and the Beryls

In the same way that ruby and sapphire are colour varieties of the mineral corundum, so emerald and aquamarine are varieties of the mineral beryl. While the beryl gemstones are less hard than the corundum ones, they are quite hard enough to wear, and very attractive – to an extent which makes production of synthetic versions commercially worth while.

Emerald is by far the most important variety of beryl and is quite plentiful. It has always been one of the most desirable gemstones and this respect is still apparent in the trade, even after many years of colour enhancement of different kinds – which have, in fact, taken place since emerald first became known as a gemstone.

Few if any emerald crystals reach a piece of jewellery without having been 'improved' in some way. Since emerald is so important and expensive as a gemstone, the practice in the gem and jewellery trade today (for all emeralds of any size and importance) is to supply a certificate of place of origin which should also carry details of any treatment that may have been applied to it. This practice is also routine for ruby and blue sapphire and saleroom catalogue entries always mention any treatment detected.

Is emerald particularly suitable for treatment – or more suitable than the other gem species? Many of the finest emeralds (the larger specimens from Colombia) have quite prominent mineral inclusions and they may, in fact, combine in such a way that the incident light by which the stone is viewed is scattered with the result that the faceted stone looks glassy. It is most unusual to find an inclusion-free emerald so that over the centuries buyers and sellers of emerald have become used to the internal furniture carried by almost all specimens.

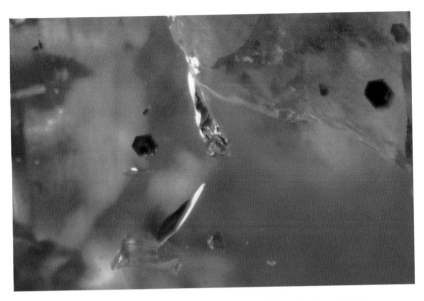

Synthetic padparadshah by Chatham contains black inclusions of platinum

Those wishing to treat emeralds with a view to minimizing the effects of solid inclusions have to ensure that the finished stone does not look unnatural. Heating the stone is not appropriate for emerald, although it is quite commonly used for ruby and blue sapphire; instead the aim is to diminish the scattering of light by the inclusions, which makes them easier to see. The appropriate treatment is to fill surface-reaching fractures with oil, a transparent glassy or polymeric (plastic) substance.

While filling of some kind has been done for centuries the production of emerald by man did not become established until the late nineteenth century and experiments had taken place long before then. Recognizing the earlier work of Ebelmen, who reported successful emerald growth in 1848, the first report of successful synthesis was made in 1888 by Hautefeuille and Perrey, who dissolved 18.75 g of the constituents of beryl in a 92 g flux of lithium molybdate in a platinum crucible. Chromic oxide (0.6 per cent) was added to give the green colour. The flux was melted first with the furnace at a dull red; the temperature was then raised over twenty-four hours to 800°C and this temperature was maintained for five days. The end product was about fifteen small crystals. A heating period of fourteen days which was attempted later gave larger crystals up to 1 mm in diameter. Today's synthetic emeralds take up to one year to grow to such a size that stones could be faceted from them.

PROPERTIES

Before looking more closely at the growth of crystals we need to know something of beryl's composition. While diamond is an element (carbon) and corundum a simple oxide, beryl is one of the many silicates, a mineral family which provides a number of gem species. The composition is beryllium aluminium silicate, $Be_3Al_2(SiO_3)_6$. It is clear that this is a much more complicated composition; each element will have its own individual qualities of melting and cooling. If emerald is grown chromic oxide has to be added to give the colour. Some natural and synthetic emeralds contain vanadium oxide and traces of iron are usually present in the natural material, stones from some localities being notably iron rich. Some vanadium-coloured beryl is certainly an emerald green but the international gem trade dictates that green beryl may be called emerald only if chromium is present to give the colour.

Natural emeralds contain some water and show the infra-red spectrum of water. Stones grown by the hydrothermal process described below also show this spectrum. Stones grown by the flux-melt method (the commonest method of growth for emerald) contain no water.

SYNTHESIS

Early Attempts

Nassau, in *Gems & Gemology*, gives an excellent account of early attempts at emerald growth, which shows some of the problems encountered by growers in the nineteenth century. Ebelmen reported that he had obtained crystals of emerald in 1848 by heating natural emerald in powder form with molten boric acid as a flux. The powder dissolved in the molten boric acid and recrystallized in very small hexagonal crystals. This is one of the first examples of flux-melt growth.

Hautefeuille and Perrey published reports in 1888 and 1900 in which they described the growth of emerald crystals using different fluxes until they found that lithium oxide with molybdenum oxide or vanadium oxide gave the best crystals, some of them reaching 1 mm across after growth at 800°C for fourteen days. When iron was present the crystals were a greenish-yellow, chromium giving the emerald colour. It is interesting to note that the beryllium silicate phenakite grew instead of beryl when the temperature exceeded 800°C – this still happens occasionally in the modern flux growth of emerald.

During the early years of the twentieth century the German firm of IG-Farbenindustrie AG (formerly Elektrochemische Werke) worked on the synthesis of emerald, although full details are hard to obtain. Nassau in *Gems Made by Man*, quotes data from some papers written by H. Espig, who was engaged in this work from 1924 until 1942. The most successful flux turned out to be lithium molybdate, which produced crystals up to 2 cm long in twelve months. It was reported that profuse inclusions made the crystals unsuitable for the faceting of stones larger than 1 ct. The inclusions were presumably of flux.

It is interesting to note that Espig stated that chromium on its own gave a bluish-green colour which he felt would not be appropriate for emerald of commercial quality and that another element was needed (perhaps vanadium, but the element was never named).

Nassau notes that the pattern of inclusions in the product, which under the name of Igmerald was never released to the trade but given in presentation boxes to a few individuals, accorded with those seen later in the productions of Nacken, Chatham, Zerfass and Kyocera. The name Zerfass emerald was given to the equally rare crystals and faceted stones grown by a colleague of Espig at IG-Farbenindustrie from about 1964, the name being taken from the firm of Walter Zerfass in Idar-Oberstein, Germany. By 1970 Zerfass emeralds were distinctly rare; while faceted stones and crystals show characteristic flux inclusions in notably hexagonal patterns the colour is very fine. Some crystals show crystals of beryl protruding from the prism faces – this makes them highly collectable and in fact there are collectors of synthetic products, many of which are rarer than most natural stones.

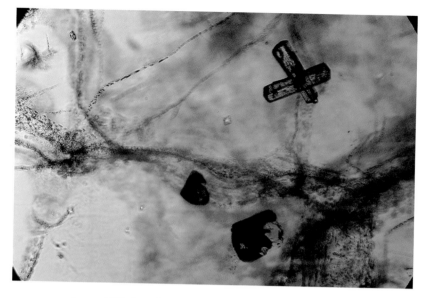

*'Igmerald'. Veils and wisps accompanied by phenakite crystals
(two in a cross-like position)*

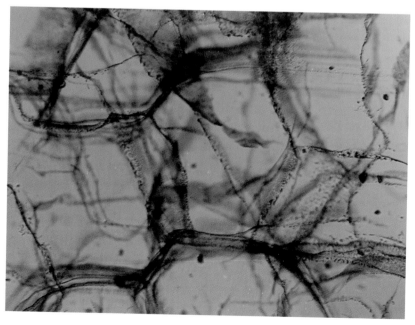

Zerfass emerald. Typical veil inclusions form a honeycomb pattern

Nacken and Zerfass emerald. Cuneiform two-phase inclusions in parallel alignment, starting from tiny phenakite crystals

Flux-grown Emeralds

NACKEN EMERALD

In the 1970s while I was in Germany, I was able to talk to the doyen at that time of pre-war German gemmology, Herr Georg O. Wild. He showed me many birds and animals carved from hardstones (they were his speciality) but also a few small, green emerald-like crystals. He explained that these were some of the Nacken emeralds, which must have been grown at some time between 1916 and 1928. I have never seen a faceted specimen, although they could exist – the crystals were just about large enough. Herr Wild was kind enough to give me a crystal.

I mention this because it throws some light on some of the problems which used to cause uncertainty and discussion, what is the method of growth of a particular crystal, has the literature been read carefully enough – and was there any published literature anyway? For many years the Nacken emerald crystals were believed to have been grown by the hydrothermal method, but there is no sign of the infra-red spectrum

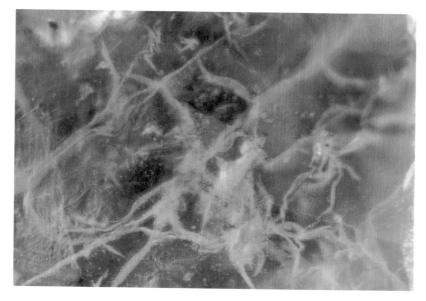

Synthetic emerald by Nacken contains wisps and veils of flux residues

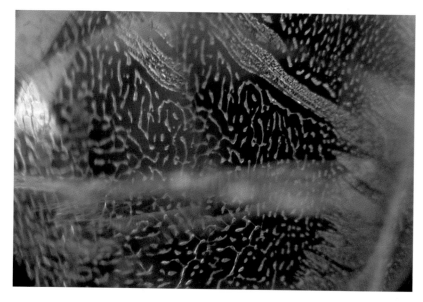

'Igmerald'. Typical and conclusive pattern of veil-like inclusions

of water which is seen in all hydrothermal products and in natural emeralds. Flux particles in which traces of vanadium oxide have been found are characteristic of the Nacken crystals.

Nassau, in *Gems Made by Man*, described some of the Nacken emerald crystals in the collections of the Natural History Museum, London. They had a recognizable prismatic form, with some showing perfectly developed basal pinacoids accompanied by adjacent curved surfaces showing fine striations. The crystals showed no signs of having been attached to any part of the crucible furniture and contained fragments of natural beryl on which seeds the emerald crystals had been grown. Some of the most interesting features of the Nacken emerald (present in my own specimen) are cuneiform nail-like inclusions formed by the nucleation of small crystals of phenakite followed by a flux inclusion and a tapered closure. In general the flux inclusions in the Nacken emeralds resemble those found in the later Chatham and Gilson products.

CHATHAM EMERALDS

Carroll F. Chatham is reported by Nassau to have grown his first emerald crystal just before the Second World War. One of the first problems was nomenclature, as the Jewellers' Vigilance Committee had objected to his original name of Chatham cultured emerald. The name was therefore altered to Chatham created emerald in 1963 with the permission of the United States Federal Trade Commission. For many years the Chatham product was the only synthetic emerald on the market and it is still offered today.

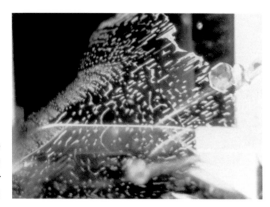

Chatham emerald. Tell-tale fingerprint veil in contact with a phenakite crystal (under dark field illumination)

Chatham emerald. As on page 167 but photographed in bright field illumination

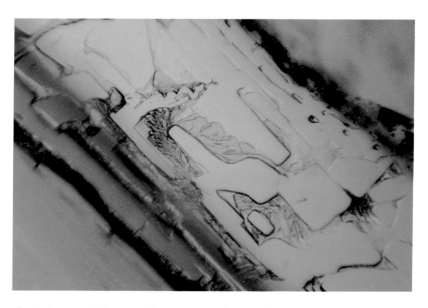

Synthetic emerald from a melt. Strong magnification of a section of one of the typical wispy veils. Note the solid impurities in the drops and tubes

It is available as single crystals and as crystal groups (most natural emeralds do not occur in similar groups). The growth period may be up to one year, the usual time for flux-grown emerald crystals. The product shows characteristic twisted veils of flux, no sign of the presence of water and the lower physical and optical properties associated with most synthetic emeralds: RI 1.56–1.58, birefringence near 0.006, and SG 2.66–2.69. Some of the RIs overlap with those of some natural emeralds and no emerald should be diagnosed as synthetic on physical and optical properties alone. The microscope should always be used and while simple instruments like the Chelsea colour filter usually give a clue to natural/synthetic origin (synthetic emeralds usually show a notably bright red) there are always likely to be exceptions. Growers are always free to add iron, which diminishes the bright red effect.

As well as the usual twisted veils of flux particles Chatham emeralds may contain crystals of phenakite. Even these are not peculiar to the Chatham product: fragments of crucible material are also common in other flux-grown emeralds.

Nassau makes the interesting point that Chatham had access only to the published reports of the work carried out by Hautefeuille and Perrey in 1888 and 1900 when he began to grow emerald crystals himself. Reports on the IG Farben emeralds and how they were grown were restricted by security considerations until 1960.

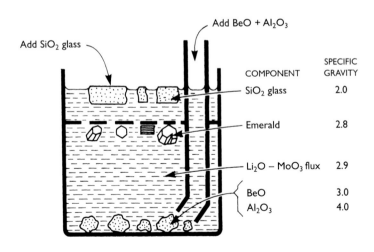

Fig. 9 The IG-Farben process for the flux-melt growth of emerald

GILSON EMERALD

In 1964 Pierre Gilson, whose firm in the Pas de Calais, France, specialized in the manufacture of ceramic materials, was able to release his first commercial emeralds onto the market. Gilson grew his emerald crystals from a flux of lithium molybdate, using colourless seeds of natural beryl. These were coated with emerald in a preliminary process and then used as seeds again once the colourless sections had been removed from the emerald ones. The final growth of emerald on these seeds took several months before a crystal large enough for fashioning was grown.

Textbooks all record that at one stage during the years of Gilson emerald production some iron was added with a view to diminishing the strong red fluorescence seen with crossed filters or the Chelsea filter. These crystals were called 'N' series, but production was not maintained. Gemmologists looked for an absorption line at 427 nm, which is not usually found in natural emerald.

Like Chatham, Gilson grew groups or clusters of emerald crystals and some of them sold for high prices. Gilson's faceted emeralds were graded before sale by a system of stars, four being high quality, with

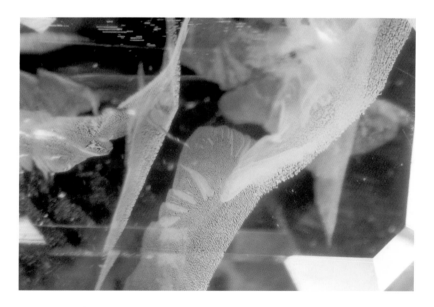

Gilson emerald. Typical veil inclusions forming 'flying' banners

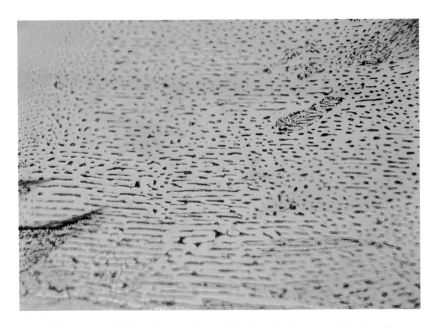

Gilson emerald. Note the conspicuous directional orientation of drops and 'hoses' forming a large veil

specimens almost clear – they were suspicious on that account, although the colour was very fine indeed. The Gilson process was sold to Nakazumi Earth Crystals Corporation of Japan and I have not heard of any production for a number of years.

LENNIX EMERALD

In 1966 L. Lens of Cannes produced the first crystals of what was later to be known as the Lennix emerald. First to be grown were small clusters but crystals large enough for stones to be faceted from them were available by the early 1980s. Lens has said that the crystals were grown by the flux-melt process with temperatures around 1,000°C and at atmospheric pressure. GIA, in a report of 1987, described the crystals as hexagonal, with a tabular habit. The predominating form was the pinacoid (in beryl the face at right angles to the vertical crystal axis). A faceted stone of 1.30 ct tested at the same time showed a homogeneous dark green colour. Clarity varied considerably with some specimens being quite clear and others heavily included by flux particles.

The Lennix emeralds showed an RI in the range 1.556–1.568 with a birefringence of 0.003 (quite low for emerald), although it turned out that the darker green areas of the Lennix emeralds showed a higher RI. The SG was in the characteristic range of emerald, 2.65–2.66.

Other features of synthetic flux-grown emerald, a bright red under the Chelsea colour filter, red fluorescence under LWUV with a weaker but similar colour under SWUV and a chromium absorption spectrum all make detection reasonably simple. One apparently unique feature of the Lennix emeralds is their response to cathodoluminescence, when some specimens show a purple or bright violet-blue response which is apparently unique among synthetic emeralds.

Lennix emeralds show the characteristic flux inclusions seen in other products; GIA noted opaque tube-like inclusions preferentially aligned parallel to the vertical crystal axis – clusters of inclusions seemed to prefer the borders of successive growth zones following the edges of the basal pinacoid. Thin crystals of beryl and phenakite, with healed fractures lined with flux, were also found. Some specimens contained an opaque black material originating probably from a molybdate flux.

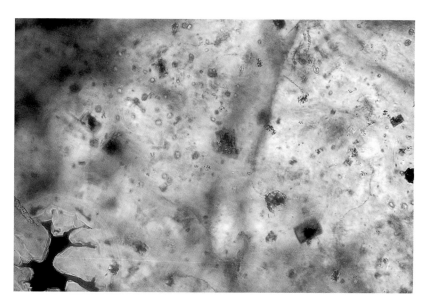

Lennix emerald. Survey of affiliating inclusion scenery. Note the strongly dichroic, green, guest crystals and the curious splashes

JAPANESE SYNTHETIC EMERALD

The Japanese firm Kyoto Ceramics (Kyocera) produced a flux-grown emerald which was available only when set in a jewellery range distributed by the firm. The emerald was given the trade name Crescent Vert (the name Inamori created emerald was used in the United States).

Another Japanese product, the Seiko emerald, showed a refractive index of 1.560–1.564 with a specific gravity of 2.65 – these figures are in the customary synthetic emerald range. The Seiko product contained planes of radiating phenakite crystals set in groups between growth layers and also single crystals of phenakite. Flux particles of a nearly rectangular shape appear to lie in a single direction in a plane between the colour zones.

RUSSIAN FLUX-GROWN EMERALD

Synthetic emeralds grown in Russia were being described by the early 1980s. A specimen described in 1985 was a cluster of self-nucleated hexagonal prisms reminiscent of the crystal groups grown by Chatham and Gilson. Like these two products, the Russian emeralds showed pinacoidal terminations with a stepped effect. The crystals in one sample at least were radiating from a crust of polycrystalline material; they reached up to 3 cm in length and 4.2 cm in diameter.

Tests carried out by GIA on a single cluster and eighteen faceted stones showed that their RI and SG, at 1.559–1.563 and 2.65 respectively, were sufficient to distinguish them from natural emeralds. The birefringence was 0.004.

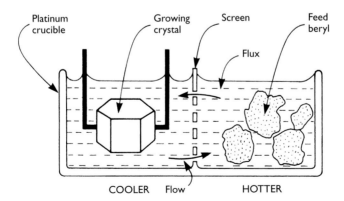

Fig. 10 The Gilson process for the flux-melt growth of emerald

But these properties, are not always enough on their own to identify the stones as synthetic rather than natural. Under magnification the surface of the cluster showed three distinct solid phases in addition to the emerald. One was phenakite, and there were single crystals and groups of alexandrite and silvery metallic platelets emanating from the lining of the growth crucible.

Two distinct forms of flux inclusion were reported. One was in the shape of secondary healed fractures while others appeared as primary void fillings. Some inclusions showed two phases with a glassy bubble while others showed fingerprint-like patterns. These were noted especially in the faceted stones. Greyish fragments of platinum have also been reported.

Hydrothermal Emerald

While the majority of synthetic emeralds are still grown by the flux-melt method, many are grown hydrothermally by a process employing a pressure vessel, similar to that used for the growth of quartz, which takes place on a large scale. Emerald crystals are grown on seeds at

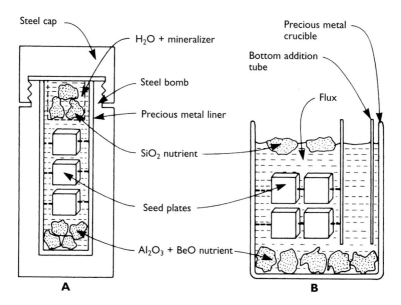

Fig. 11 Making synthetic emerald. This compares hydrothermal growth (A) and flux-melt growth (B)

typical temperatures of 500–600°C and pressures of 700–1400 bar. Growth rates are in the region of 0.3 mm per day.

LECHLEITNER EMERALDS

The work of the Austrian grower Johann Lechleitner seems to have appeared on the market in the early 1960s. It is not easy to keep up with the various products he has grown but a most useful paper published by *Australian Gemmologist*, vol 17, no. 12, 1991, gives a summary.

Type A emeralds consist of flux-growth emerald on a seed plate on natural beryl. The plates were cut in the direction of the C-axis (vertical crystal axis). They were grown between 1958 and 1959 and sent only to gemmological laboratories.

Type B emeralds were placed on the market. They had cores of colourless or slightly coloured greenish beryl which were covered by an emerald layer grown hydrothermally. Both Fe^{2+} and Fe^{3+} ions were detected in the core, in which chromium was also present. The overgrowth showed fissures and cracks.

Type C emeralds also used seeds of natural beryl cut obliquely to the C-axis. They were said to have been grown in a single run. The seed

Lechleitner emerald. The surface displays a dense pattern of fine fissures

shows dark green layers of emerald on both sides of a colourless centre. None of these specimens reached the market: growth took place between 1962 and 1963.

Type D emeralds are complete rather than overgrowth on a seed which is taken from the hydrothermal emerald layers of Type C material. These were first grown in 1964 and show a layered structure under magnification. Type E emeralds were grown for research purposes only and Type F ones were grown between about 1972 and the 1980s.

At the time of writing only Types B, D and F emeralds are available. In general the Lechleitner products can be recognized by fissures and cracks: the stones grown on seeds often show a 'crazy paving' effect.

The earliest Lechleitner emeralds grown on to faceted natural beryl seeds were sometimes fashioned with the back facets left unpolished. Marketed by the Linde firm (see below) and by a local firm in Austria, the trade names originally used included Symerald and Emerita. The name Emeraldolite has been given to a green material consisting of flux-grown emerald grown on a seed of opaque white beryl. This product was made in France and echoed Lechleitner stones in some ways, although they were flux rather than hydrothermally grown.

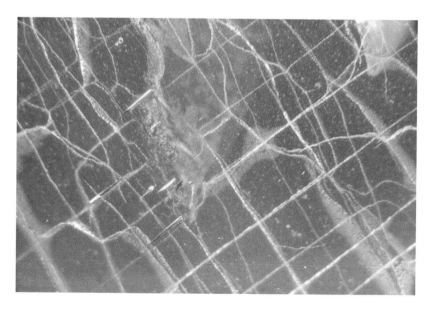

Lechleitner emerald. Strong magnification displays the nature of the fissures, which form a narrowly meshed net in the synthetic mantle

Lechleitner 'Emerita' emerald. Two systems of net-like fissures meet along the facet edges of the pre-cut core

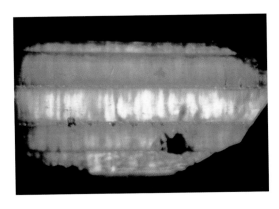

Lechleitner 'sandwich' emerald. Lateral view of the central colourless beryl and the synthetic crown and pavilion, which impart green colour to the whole

Emeraldolite is opaque and the coating is only 0.3–1mm thick. Growth takes place differentially so that only asymmetrical shapes result and for this reason cabochons are not fashioned. They could not, because of their lumpy appearance, be offered as natural emerald. This limitation seems to affect all emerald growth.

Emeraldolite has a mean RI of 1.56 and an SG of 2.66. Stones show a chromium absorption spectrum and appear brownish-red through the

Chelsea filter. The material is tough and hard (over 8); the overgrowth cannot be broken away from the core. The flux layer under magnification was found to show groups of minute parallel crystal faces. White beryl showed through gaps in the overgrowth. Emerald crystals in the overgrowth showed characteristic faces for beryl. The 'crazy paving' effect seen with some Lechleitner stones was not apparent. Large flux inclusions could be seen with spherical voids resembling the gas bubbles seen in glass.

LINDE EMERALDS

From 1965 the Linde Division of Union Carbide Corporation grew emeralds by the hydrothermal process. By 1969–70, 200,000 ct were produced

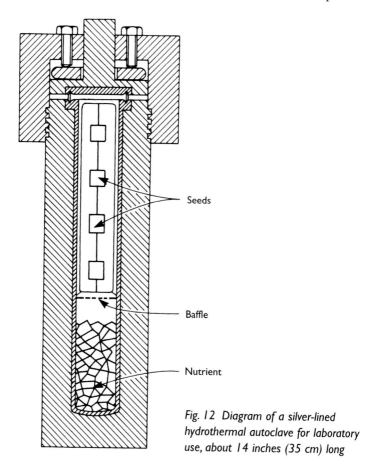

Seeds

Baffle

Nutrient

Fig. 12 Diagram of a silver-lined hydrothermal autoclave for laboratory use, about 14 inches (35 cm) long

annually. The growth rate reached 0.8 mm a day. As happened with the Japanese Seiko emerald the Linde stones were at first available only in the company's own manufactured jewellery, which was marketed under the trade name Quintessa. I visited the growth facility in the early 1970s and examined a number of crystals.

The Linde emeralds (later Regency) showed an RI of 1.566–1.578, with a birefringence of 0.005–0.006 and an SG of 2.67–2.69; these values corresponded to those that might be expected from natural emeralds. As usual with hydrothermally grown materials there are few inclusions compared with flux-grown specimens and there are no solid inclusions – always the best feature of hydrothermal synthetic stones. Nevertheless what can be seen are crystals of phenakite and very delicate two-phase inclusions. Stones give a notably strong red flash when viewed through the Chelsea filter. Very small bread-crumb-like inclusions have been reported and chevron-like markings appear to be a clear indication of hydrothermal growth. There are no liquid or liquid-like inclusions.

Linde hydrothermal emerald. Thin seed crystal. On either side, cuneiform two-phase inclusions dart out at an oblique angle to the surface of the seed plate

BIRON EMERALD

A grower in Australia began the synthesis of emerald by the hydrothermal method in 1977, selling the stones under the trade name Biron. Before looking more closely at the stones it is worth mentioning that the Pool and Kimberley emeralds, also grown in Australia by the hydrothermal method, have features which appear to show that they could easily be the Biron product under another name. Pool emerald was at one time advertised as 'recrystallized natural emerald' but the Biron trade name was in fact restored to the product in 1988. A little nomenclature confusion.

From the gemmological point of view the Biron emerald's properties more or less overlap with those of natural emerald, if slightly lower than some, with an RI of 1.569–1.573, a birefringence of 0.004–0.006 and an SG of 2.68–2.71. Specimens were inert to UV radiation even though iron was not found to be present. It is possible that the non-response is due to the presence of vanadium. Many natural emeralds do not respond to UV, however.

Under the microscope the Biron emeralds show some unique features. Some metallic gold-like fragments proved actually to be gold, and probably came from the autoclave lining. On magnification the fragments were sometimes angular grains or thin plates. Well-formed crystals of phenakite have also been found and might be confused with the clear, colourless calcite found in some natural emeralds. If the inclusion is large enough one might see brighter interference colours in calcite than in phenakite when viewed between crossed polars. Some experience is needed for this and a polarizing microscope is the best tool. The phenakite inclusions in the synthetic product are found in greater profusion than calcite crystals in the natural emerald.

An effect known as a 'Venetian blind' has been seen in the Biron emerald (something similar can be seen in the rare Verneuil-grown red spinel). Neither this nor prominent colour zoning is a feature of natural emerald and an effect resembling comet tails (whitish particles, seen in some synthetic rubies) is not known from other synthetic emeralds. Some reported specimens have shown near colourless seed plates flanked by planes of gold inclusions.

Many hydrothermal emeralds contain cone-shaped voids resembling spicules, with heads like those of large rectangular nails, and these can be seen in the Biron products. The nail-head is a single phenakite crystal or a group of them; the voids contain fluid and a gas bubble. Biron emeralds contain particularly deceptive veils of flux but natural emeralds do not show the type of fingerprint liquid inclusions common in corundum. Biron emeralds have been found to contain chlorine, which has not been reported from natural nor from flux-grown emeralds.

RUSSIAN HYDROTHERMAL EMERALD

Emerald grown by the Laboratory for Hydrothermal Growth of the Russian Academy of Sciences, Novosibirsk, were described by GIA in 1966. The RI was in the range 1.572–1.584 with a birefringence of 0.006–0.007 and an SG of 2.67–2.73. These values overlap with those of natural and other synthetic emeralds. Of the eight stones tested none showed any response to either type of UV radiation. Through the Chelsea filter a weak red glow could be seen when the stone was held at low angle to the lamp. No useful information could be gained either with the dichroscope or spectroscope.

The chevron-like markings characteristic of other hydrothermal emeralds could not be seen in the Russian stones. Reddish-brown particles of unknown origin seemed to be arranged in dense clouds. Using fibre-optic sources clouds and layers of small white particles were seen in all the specimens examined. A phenakite crystal was seen in one of the stones. The Russian stones seem to be character-ized by the reddish-brown and white particles. Traces of nickel, copper and water have been established and stones also show sets of parallel lines in a step-like formation, which arises from reportedly fast growth. The step-like structures are parallel to the surface of the seed.

Faceted specimens examined showed parallel growth planes at about 45° to the optic axis. Seeds had been cut parallel to a face of the second-order hexagonal dipyramid. This orientation avoided the growth patterns seen in Russian hydrothermal emerald grown previ-ously. Growth was reported to have taken place in steel autoclaves without noble metal linings.

Above and below Russian synthetic emeralds containing the usual signs of man-made growth and absence of natural inclusions

Russian synthetic emeralds containing the usual signs of man-made growth and absence of natural inclusions

Other Synthetic Beryls

AQUAMARINE

The great value of emerald probably makes it worth 'growing one's own', but this is not necessarily true of the other colour varieties of beryl. Aquamarine is so easily simulated by glass and by synthetic spinel grown at very low cost that synthesis would hardly seem worthwhile. None the less aquamarine has been grown in Russia using the hydrothermal method; the growers were at the Institute of Geology and Geophysics, the Siberian branch of the Russian Academy of Sciences at Novosibirsk.

The blue colour is caused by iron in both ferrous and ferric states and crystals have an RI of 1.575–1.583, with a birefringence of 0.008 and an SG of 2.69, which are consistent with natural aquamarine. Iron and nickel can be detected from visible and ultra-violet absorption spectra and the infra-red spectrum shows that water is present. Under magnification the junction between the near-colourless seed and the blue overgrowth is not hard to see and a very weak colour zoning can be

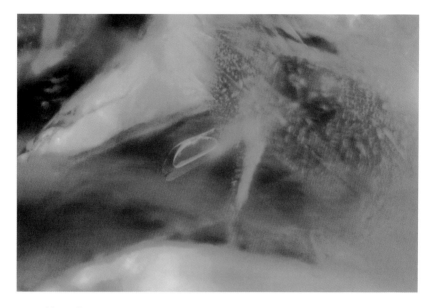

Natural aquamarine may appear inclusion-free but signs of artificial origin are apparent here

Mauve beryl is scarcely known in nature and any example such as this needs careful attention. The metallic crucible fragment is a clue

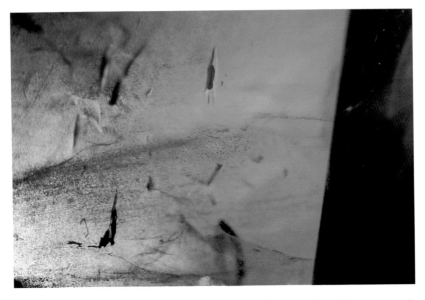

Red synthetic beryl is presumably grown to imitate ruby rather than the rare natural material. The two-phase inclusion gives away its origin

seen in the overgrowth in a direction parallel to the boundary with the seed plate. As the inclusions in both seed and overgrowth are similar it would seem that the seed is also hydrothermally grown. Seed and overgrowth show a cellular structure seen best when the specimen is viewed, immersed, at about 60x magnification. Cavities with multi-phase fillings have also been observed, along with different structures resembling feathers. One type is flat and the other twisted; they contain trapped shreds of growth solution.

The cellular and other structures provide the best means of identification. Specimens have also been found to show the absorption spectrum of nickel (the hand spectroscope will not distinguish this), which has not been reported from natural beryl. The origin of the nickel is the wall of the pressure vessel, which for this growth does not have a precious metal inner lining. Amounts of sodium and magnesium, together with a high content of iron, show that this aquamarine cannot have arisen naturally.

The structure of beryl is able to accommodate water in a central channel between the rings forming the crystal. Infra-red spectroscopy is able to identify the absorption spectrum of water (it cannot be detected by the direct-vision spectroscope, which of course operates using visible light). Flux-grown emerald crystals contain no water so that if the water spectrum is found the specimen must be hydrothermally grown or natural. The microscope should have no difficulty in resolving the mineral inclusions, almost invariably present in natural crystals.

VANADIUM 'EMERALD'

The part played by vanadium in the coloration of green beryl is of interest to gemmologists since the 1960s when A.M. Taylor of the Crystals Research Company of Melbourne, Australia, grew an emerald-like green beryl devoid of chromium.

At the time of the original report crystals were said to have reached 10 ct, giving faceted stones up 2 ct. The RI was 1.566–1.575 with birefringence about 0.005. Marked colour banding was seen when specimens from earlier growth runs were immersed, although this effect was diminished in later products. The SG was in the emerald range at 2.68. Stones showed no response to either type of UV radiation nor was there any response to X-rays. The colour is reported to be a warm grass-green with a hint of yellow; stones were less blue-green than some other emeralds. The dichroism was yellow-green: green with the ordinary ray giving a green colour and the extraordinary ray pink. Inside the stones examined were traces of the seed placed at an angle to the C-axis. There were no reported solid inclusions.

MAXIXE-TYPE BERYL

Every few years Maxixe-type beryl hits the market again. This material is described in most of the gemmological textbooks and is probably best known for its fine dark blue colour and its propensity towards fading.

The distinction between Maxixe and Maxixe-type beryl is that Maxixe is used for naturally occurring blue beryl while Maxixe-type is used for a beryl of another colour (usually pink) which has gained its blue colour through treatment.

In the Fall 2001 issue of *Gems & Gemology* a green-blue beryl of the Maxixe-type is described. Two faceted oval stones weighing 30.67 and

23.64 ct showed properties consistent with beryl: RI 1.572–1.580, SG 2.71, and a uniaxial optic sign. When the stones were viewed face up parallel to the optic axis a strong green-blue colour could be seen. Viewed in a direction at right angles to this the colour was a greenish-yellow.

The stones showed yellow through the Chelsea filter and were inert to both forms of UV. Long needles and growth tubes could be seen inside the specimens, some extending the whole depth of the stone. Internal growth zoning was also observed and there was no doubt that the stones were of natural origin.

The strong dichroism of green-blue/greenish-yellow would be most unusual in a natural, untreated green to blue beryl. Since the darker colour was visible along the direction of the optic axis (the direction of the ordinary ray) and the lighter colour in a direction at right angles to this the identification with Maxixe beryl was complete. Natural aquamarine's dichroic colours appear in the reverse way.

When the stones were examined with the hand spectroscope the normal aquamarine absorption line at 437 nm was not apparent. On the other hand a series of bands could be seen: these were found in the red to the green sections of the spectrum, an effect that has been seen in Maxixe beryls.

Cesium was detected in both of the stones. This element has been reported from both Maxixe and Maxixe-type beryls in a number of reports. In the two stones small amounts of iron were found together with traces of manganese.

Another stone examined by GIA at a different time was a dark violet-blue colour. This also proved to be Maxixe-type. They report that five stones of this kind had been examined in an eight-month period which seems to suggest that a supply is being leaked onto the market.

OTHER COLOURS

Apart from emerald and aquamarine, no other colours of beryl have been synthesized, save for a few more or less experimental products – some of which have got onto the market for a short time. One is a water-melon beryl with a pink core and green rind grown by the Adachi Shin Industrial Company of Osaka, Japan. I have seen this rare product only in the form of a hexagonal crystal. The RI was reported as 1.559 and 1.564 with an SG of 2.66. It was reported that the crystals were grown by a method in which fluorine and oxygen reacted at temperatures higher than those obtaining in normal beryl growth with crystalline or amorphous beryllium oxide, silica and alumina (the constituents of beryl). Chromium for the green and manganese for the red were the most likely dopants. The melt after heating migrated to a cooler part of the apparatus and then settled on seeds. The company has said that crystals up to 1 cm have been grown in brown, pink, reddish-brown, colourless, sky blue, yellowish-green and emerald colours.

The hydrothermal process has been used to grow a red beryl, so far for experimental purposes only, though I have seen specimens.

Production and Testing of Synthetic Beryls

In *Gems Made by Man*, Nassau reported that in 1956 world production of synthetic emerald was approximately 50,000 ct and that estimated production for 1980 was more than ten times that amount. The majority of synthetic emeralds were grown by the flux-melt method. He also gives a brief overview of other attempts to grow emerald, making the point that emerald does not easily form crystals from the pure melt, tending rather to take a glassy structure. He shows that even when crystallization is achieved at the melting temperature it will melt incongruently, forming a mixture of emerald with another compound so that pure emerald cannot be obtained simply.

Another difficulty in the way of the would-be emerald grower is the high toxicity of some beryllium compounds soluble in water or biological fluids. In beryl and chrysoberyl the beryllium, fortunately, is too strongly bonded with other substances for it to be toxic.

It should be said that testing an emerald-like unknown for its natural or artificial origin cannot always be done by the normal gemmological tests. Considerable familiarity with the internal furniture of both natural and synthetic emeralds is needed and this means that one has to become microscope adept. It cannot be emphasized too strongly that the gemmologist must continually consult sources of information; with the Internet this is now much easier. It is likely, given the relative frequency of fine large emeralds turning up in commerce, that recourse to legal expertise will be commoner with beryl than with corundum, so that any large apparently clear stone should always be regarded with suspicion.

NOTES FROM THE LITERATURE

In 2001 *Australian Gemmologist* 21(2), gave a short account of an emerald grown in China by the hydrothermal method. Five crystals were examined; their colour was a slightly bluish-green and they showed a strong red through the Chelsea filter. Under LWUV they gave a moderately strong red fluorescence. Using the electron micro-probe it was found that the specimens recorded a notably lower level of chlorine and higher alkali than other hydrothermal emeralds (0.06–0.25 wt% chlorine and 1.09–1.30 wt% sodium oxide). The RI was measured at 1.569–1.571 for the extraordinary ray and 1.572–1.578 for the ordinary ray. The SG ranged from 2.68 to 2.73.

Inside the specimens two-phase inclusions could be seen, consisting of liquid and gas phases. The infra-red spectra in the 3,500–3,800 cm-1 region showed only a single peak at 3,701 cm^{-1} but in the 5,000–5,500 cm^{-1} three strong peaks could be seen. This is characteristic of synthetic emeralds as the natural stones show only two peaks.

In the journal *X-ray Spectrometry*, 29(2), 2000, fifty-six natural emer-alds from Colombia, Pakistan, Zambia and Brazil and twenty-six synthetic emeralds from eight manufacturers, using both hydrothermal and flux-melt growth methods, were tested using micro-proton-induced X-ray emission spectra (micro-PIXE) techniques.

Five flux-grown emeralds manufactured by Chatham and one Lechleitner hydrothermal sample gave nearly featureless data which did not occur with the natural stones. In the natural emeralds colour zoning was found to correlate with concentrations of chromium, van-adium and iron, although one Colombian specimen did not fit into this pattern.

Chlorine was found in both natural and the hydrothermal emeralds.

The green chrome mica was known as fuchsite until the revision of the mica group nomenclature by a subcommittee of the Commission on New Minerals and Mineral Names of the International Mineralogical Association in 1998. I mention this now in case any reader consulting a current mineralogical text may fail to find the name. None the less green mica with chromium-caused colour does still exist. In the Summer 2002 issue of *Gems & Gemology* a massive green mica imitation of emerald was described, the specimens being three translucent cabochons sent to the Gem Testing Laboratory in Jaipur, India.

The SG of the three stones measured together (they were small) was 2.89 and the approximate RI 1.58. Raman investigation of five spots showing green gave values for the potassium mica, muscovite. EDXRF showed that chromium was present. The point of including this example is the unusual presence of oil impregnation in a mica. The flat polished base of one of the cabochons showed a higher saturation of green on the rim and oil (possibly Joban oil) was confirmed by FTIR spectroscopy.

It had previously been suggested that there was a correlation between chromium and chlorine in synthetic emeralds and this might have been due to the use of chromium chloride (6 water) to provide the chromium to give the colour. This relationship did not feature in the study.

Mineral inclusions of calcite, dolomite, chromite, magnetite, pyrite and several mica and feldspar group species were identified in the natural emeralds. The association of the two forms of (zinc, iron) sulphide (sphalerite and wurtzite) with pyrite were noted for Zambian emeralds. The authors of the paper concluded that this method of analysis was of considerable potential value for the study of emerald.

Gemmology Queensland reported in June 2003 a new emerald-green synthetic material marketed through Rough Synthetic Stones of Bangkok (the original report appeared in *Jewellery News Asia*, May 2003). The stones were sold under the name Fortall.

The material was said to be grown by crystal pulling and had a hardness of 7, RI 1.72 (the material was believed to be singly refractive) and an SG of 2.70.

As well as the emerald-green variety, black, blue, brown and red colours were said to be available as either rough or cut specimens. It was first used industrially.

CHAPTER EIGHT

Opal

PROPERTIES

Opal is silica with a variable amount of water; the formula is written $SiO_2 + nH_2O$. The SG is near 2.10 and the RI about 1.45. The water content is variable as the formula shows; some opal may crack or craze when mined and the onset of this process is impossible to predict though some areas seem to produce specimens which have not altered over years. The crazing cannot be halted.

Opal's considerable attraction arises from the play of colour against a light or dark background (called white and black opal respectively, whether or not the background is really white or black).

Water opal is water-clear with a play of colour (in the United States this variety is called jelly opal). Transparent fire opal has a red, orange or yellow body colour and may or may not show a play of colour.

SYNTHESIS

Until the cause of opal's play of colour was at last worked out in the 1950s it was possible only to make imitations rather than a true synthetic product. It is now known that the effect (not opalescence – this refers to the milky effect visible in many opals) is caused by diffraction of white light by a regular three-dimensional array of equal-sized amorphous, water-bearing silica spheres and voids. Light rays scattered from the array interfere with one another and produce all possible colours, the size of the spheres determining which colour is produced.

Successful synthesis was first reported in the second half of the twentieth century and from that time a large number of materials (including latex) have been used to form arrays but the problems of strength and durability have shown that silica is most effective for ornamental use. Full details of the growth processes used are not easily available.

None the less, Nassau reported in *Gems Made by Man* that monodisperse silica spheres are prepared first. They can be formed from the organic silicon compound tetraethyl orthosilicate. This can be dispersed as fine droplets in a mixture of alcohol and water. The addition of ammonia or other mild alkali converts the droplets into silica spheres which will contain a small amount of water. Close control of stirring and concentration is critical, the materials used must be pure and the spheres must be the same size.

The spheres are allowed to settle in a water of controlled acidity, a process that can take at least a year. The product of this process needs care as it can easily dry out and lose its colour as a result. Some sintering and mild pressure can be used and it is also possible to fill the pores between the spheres with additional silica.

Gilson Opal

The properties of synthetic opal are very close to those of the natural material, although magnification almost always shows a highly characteristic hexagonal patterning within the colour patches. A pioneer in synthetic opal production was Pierre Gilson, who is well known as a grower of synthetic emerald and other ornamental materials. His product was known as Gilson created opal and both black and white types were produced. The opal was made up of columns which extended vertically from the base of the cabochons into which the stones were cut. These structures, when viewed from the side of the stone, were seen to extend from top to bottom, an effect not seen in natural opal. If the synthetic opal was used in doublets or triplets the columns appeared to stop suddenly.

Gilson has also synthesized fire opal and water opal, and they showed properties in the same range as the natural material. One Gilson fire opal with a brownish-orange colour was reported to show an excellent play of colour when examined from above. Viewed from the side the stone showed that the play of colour was confined to a central section on the top and bottom of which were thin colourless areas with no play of colour. The body colour of the specimen came from the brownish-orange central area. Markings characteristic of synthetic opal were seen in the colourless area. The different areas varied in hardness and in their response to UV. The colourless surface was easily indented with the point of a pin and also flowed when the thermal reaction tester was brought close to it. This must have been one of the many experimental products which reach gem testing laboratories from time to tome.

Russian Synthetic Opal

A synthetic opal has been made in Russia; the specimens examined showed both black and white backgrounds to the play of colour. Examination at the Gem and Mineral Show in Tucson, an area with a notably hot and dry climate, showed crazing, though examples examined previously and elsewhere did not show this effect. Examples had an unusually low SG for opal, at 1.75–1.78 (natural opal is around 2.10). An unusually high water content was thought to be responsible.

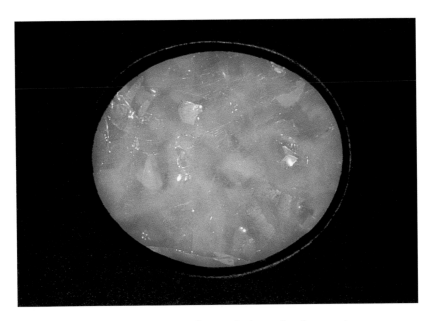

'Slocum stone'. Low magnification displays only colour patches.
Note their sharp contours

'Slocum stone'. High magnification reveals the discrete cuttings of aluminium foil which reflect interference colours

Stones gave a strong bluish-white to white fluorescence under LWUV and a greenish-yellow to blue to white reaction to SWUV. The play of colour was similar to that shown by natural opal. It is believed, on the evidence of IR spectroscopy, that some organic compounds were present – they could have played some part in sphere consolidation. Black cabochons of the Russian opal showed an RI of 1.35 and an SG of 1.65. These low figures may have some relationship to the presumed organic content.

The name Slocum stone has been given to an ingenious glass imitation of opal in which the play of colour is caused by thin scraps of metal foil. Viewed through the base the metal shows a distinct purple colour, unlike anything seen in natural opal.

Kyocera Opal

Opal has been made by the Japanese firm of Kyocera (Kyoto Ceramics) since 1980, most of the production being black and white opal. *Gems & Gemology*, Winter 2003, reported semi-transparent varieties, including fire and water (jelly opal in the USA).

Stones illustrated in the article weighed between 2.41 and 2.68 ct and were cut as oval double cabochons. They gave RIs, using the spot method, between 1.46 and 1.47 and hydrostatic weighing gave SG readings between 2.22 and 2.27. The white and near colourless specimens showed a very weak to weak chalky-blue fluorescence under LWUV with a similar but stronger response to SWUV. The black cabochon was inert to both forms of UV and none of the stones showed any phosphorescence.

With a hand spectroscope an absorption line could be seen at 580 nm, with a band at 550 nm in three of the specimens tested. The orange stone showed only general absorption to 500 nm. The near-colourless and the black cabochons showed absorption lines at 620 and 600 nm. All the stones showed the now-expected hexagonal patterning within the colour patches, a sure sign of synthetic opal.

The manufacturers stated that the opals were not polymer impregnated and FTIR spectroscopy bore out this claim. No polymer-related evidence could be found. The SG measurements on the new Kyocera opals was higher than those made on earlier polymer-impregnated stones described in the past, about 1.88–1.91.

NOTES FROM THE LITERATURE

A paper in *Australian Gemmologist*, 21(7) 2002, and abstracted in the Spring 2003 issue of *Gems & Gemology*, described the development of the method and the stages of manufacturing pure silica opal at the Centre for Applied Research, Dubna, Russia.

The first stage was the synthesis of monodisperse particles of silica in alcohol-based solutions. This was followed by the precipitation of a 'raw' opal precursor by spontaneous sedimentation or centrifuging. The third stage was the drying of the precursor opal to remove liquid from its pores. The final stage was the filling of the pores with silica gel and the sintering of the samples at 825°C. The physical, chemical and gemmological properties of this material were identical with those of natural opal. Synthesis is reported to take about ten months (the authors cite twelve months for the growth of the Gilson synthetic opal and eighteen months for an experimental Chatham product).

Sanwa Pearl Trading told *Gemmology Queensland* in 2003 that an opal imitation was being manufactured in Germany by a method involving high pressures and temperatures. No epoxy resin was involved, although a special bonding agent was used.

The material known as treated opal matrix is quite effective and rather attractive. Details of the treatment vary but the aim is to darken the background of white opal with a reasonable play of colour by incorporating minute spots of carbon. The carbon is often said to be obtained by heating old sump oil, in which the opals are steeped. Identification turns on the dark spots of carbon with which the colour patches are mingled. No untreated opal shows anything like this effect. It is always surprising that no trade name has been widely used for this material but the practice must be carried on over a wide area.

CHAPTER NINE

Quartz

PROPERTIES

The single crystal quartz (silicon dioxide, SiO_2) has an SG and RI which remain unchanged from variety to variety and from specimen to specimen. The hardness is 7, the SG 2.651 and the RI 1.544–1.553, with a birefringence of 0.009. The crypto-crystalline quartz varieties (see below) are close enough to these figures for mistakes to be avoided, and in any case their appearance is generally distinctive. Distinguishing synthetic from natural quartz cannot depend upon these constants but upon other observations described below.

It is perhaps surprising that quartz in its many forms is widely synthesized, since it is plentiful in its natural state. It is possible that for slightly cheaper lines of jewellery customers expect completely clear and strongly coloured stones, when with ruby and emerald, for example, inclusions and irregular colour distribution are some sort of guarantee of natural origin.

The quartz gemstones fall into two classes, distinguished from one other by their crystallinity. The transparent varieties are single crystals and include the colourless rock crystal, the attractive grey or brown smoky quartz, purple amethyst and citrine with colours ranging from near-red through orange to yellow and brown. The green aventurine gets its colour from profuse inclusions of a chromium-bearing member of the mica group of minerals (the name fuchsite was current before the nomenclature of the mica group species was revised).

Rose quartz, with a delicate pink colour, is found in quite large quantities in Madagascar, a major gem-producing country today, but

since the star effect (asterism) is often found in rose quartz it is sometimes synthesized.

The other class of quartz is known to mineralogists as crypto-
crystalline, made up not of a single crystal but of many, too small for
individual crystals to be distinguished by optical means, only by electron microscopes. Most if not all of these varieties will be known to the
general public under a bewildering and irritating variety of names.

It is fortunate that these materials are relatively cheap (although fine
examples of some of them can be surprisingly expensive), and they are
not synthesized. They are almost always dyed, however, but the dye is
not hard to detect with gemmological instruments and in the case of
one variety, agate, dyeing is more or less taken for granted.

The crypto-crystalline nature of agate allows highly characteristic
banding to develop as the structure is formed and the bands will
usually display different colours. The patterning can be very beautiful,
and in my opinion needs no artificial enhancement. Agates from
coastal areas of Scotland were collected in the nineteenth century and
after by such authorities as Matthew Heddle, and these specimens
show soft colours.

The Colours of Quartz

While rock crystal with no added impurity (such as iron or titanium) is
colourless, the presence of such trace elements may cause colour. Iron in
the ferrous state (Fe^{2+}) may give a rather quiet green. Transparent specimens have been called prasiolite, but the name is not widely used,
probably because the green is not exciting enough. Some prasiolite
which may have begun life as ferric iron (Fe^{3+}) gives the yellow to brown
colour of citrine. For the amethyst colour to develop the presence of an
aluminium impurity is necessary. This forms a precursor to the formation of a colour centre. In general, gemstones whose body colour is
known to fade under the action of various radiations are likely to owe
their colour to a colour centre.

SYNTHESIS

As the most important varieties of quartz, amethyst and citrine, with
rock crystal, are plentiful, certainly when compared with the major

species, it may seem odd that all are synthesized on a fairly large scale. The hydrothermal method, which is described more fully in the chapter on emerald, has for many years been used to grow often very large pure and colourless quartz crystals for use in the electronics industry, and the growth of ornamental varieties of single-crystal quartz follows the same methods, apart from producing colours.

Crystals up to 10 m in length can be grown by the hydrothermal method which, as its name suggests, uses water and heat, which together form high-pressure conditions for the crystals to grow on a prepared seed. For the production of crystals for electronic use the seed is of great importance; it needs to be cut in orientations which are of electronic significance. This is of course not so important if the crystals grown are to be for ornamental use alone, but we should remember that seed production has to be carefully controlled and that growers will be accustomed to a considerable number of precise engineering conditions. For this reason there are not many amateur quartz crystal growers. Perhaps the possibility of the pressure vessels (autoclaves) exploding (they are called 'bombs' for this reason) also discourages the tool-shed crystal grower.

Hydrothermally grown crystals of colourless quartz have nothing to tell the gemmologist about their artificial origin, or nothing much. The one difference – and this may not always be apparent – is the absence from the synthetic material of those natural mineral inclusions so common in rock crystal. Such inclusions may be prominent and ornamental in themselves. Golden acicular (needle-like) crystals of rutile and thicker dark green tourmaline crystals are most often seen in those often large specimens of rock crystal offered as ornaments.

Nothing like this is present in colourless hydrothermal quartz, so the gemmologist can be thankful that whether or not a particular specimen of rock crystal is natural or synthetic is of little importance. Most examples are not used in expensive jewellery but can be found in crystal healing shops and at gem and mineral shows.

When it comes to amethyst the picture is quite different. Natural amethyst of good quality can fetch surprisingly high prices given that specimens are not difficult to find. Stones of the finest purple colour, which used to be called 'Siberian', are easily produced by hydro-thermal growth, and apart from the absence of natural solid inclusions testing is not particularly easy. During the 1980s it was reported that synthetic amethyst accounted for up to 25 per cent of stones carried by dealers in the Far East.

Gemmological training and, in particular, experience, are necessary if one is to identify synthetic amethyst. The specimen may be placed with its main axis parallel to a ray of light which passes through the optical train of a microscope or at right angles to crossed polars in a polariscope. In this direction most natural amethyst will show the rainbow-like interference colours deriving from a form of twinning known as Brazil twinning. Most synthetic amethyst shows a succession of colour bands. I will not describe this test in detail here as interpretation of the results really is quite difficult; gemmologists should look first for traces of natural solid inclusions. Other gemmological tests will give results identical to those given by natural amethyst.

Synthetic citrine is less common than synthetic amethyst, as the public in general appears to prefer most colours to yellow (a preference I do not share). While the successful synthesis of amethyst had to wait until the cause of the purple colour was fully understood, the colour of citrine was easier to produce.

In 1973 I was able to examine what must have been an early large faceted specimen of synthetic citrine, grown by the hydrothermal method at the laboratories of Sawyer Research Projects. The stone weighed 49.28 ct and had the normal gemmological properties of quartz. No colour zoning could be seen; there were, however, very small breadcrumb-like crystals forming groups. It is believed that these crystals could be a sodium-iron silicate. It may have been formed by the reaction of iron with sodium and silica in the growth solution.

Some hydrothermal quartz displays an effect known as 'heat shimmer', which is strongest in the vicinity of the seed crystal. It arises from an area of discontinuity between the seed and the overgrowth. In the cut stone I have just described no such effect was visible but this is probably because crystals grown by the hydrothermal method can be quite large and thus contain sufficient area from which completely clear stones can be cut. One seed I examined showed one set of faces at the ends, where growth is slow, and different forms in areas where growth is more rapid.

The manufacturers of the citrine claimed that they were able to irradiate colourless quartz to give a smoky effect, using an ionizing radiation. In a green transparent quartz made by the same growers the seed area could easily be seen. This particular colour of green does not resemble that of any other gemstone commonly found.

Gemmological testing has been assisted in specimens which carry traces of the platinum wire which was used to hold the seed during growth; sometimes the wire protrudes from the crystal – an effect not found in natural quartz crystals.

A quartz crystal grown in Russia had been grown from a plate of clear quartz from which two wire loops protruded. The whole crystal showed colour variations, with colourless material on either side of the seed; then came a green band and then a brown band. This was apparently the result of experimental irradiation.

A transparent blue quartz sometimes appears on the market. Like most hydrothermally grown products cut specimens can be large, and stones I have examined over the years are transparent and a fine bright blue.

The absorption spectrum is that of cobalt and the stones show a very bright red through the Chelsea filter. It has been reported that in the absorption spectrum the central absorption band appeared to be narrower than its counterpart in the absorption spectrum of synthetic spinel doped with cobalt. A blue cobalt-doped quartz reported by GIA in 1993 gave the cobalt absorption spectrum as expected. However, when examined between crossed polars, the unique quartz interference figure was observed along the length of the main crystal axis – an uncommon phenomenon. In diffused transmitted light wedge-shaped zones of darker colour could be seen alternating with very light blue zones. These could be seen best with the unaided eye.

The type of synthetic corundum so often offered as 'alexandrite' looks much more like amethyst than alexandrite. The light in which the specimen is viewed is fairly critical but there are occasions when confusion could arise. The spectroscope will show the strong absorption band at 475 nm in the corundum imitation – no absorption bands in the visible region are seen in amethyst.

During 1978 I was able to examine two synthetic amethysts believed to have been grown in the then USSR. My descriptions were published at the time but later, in Moscow, I was able to examine some rough amethyst. The specimens at first sight could well have been natural but in one direction there was a trace of a mauve to brown colour. There was no trace of a seed or of the heat shimmer effect. Though there have been reports over the years that some synthetic quartz transmits SWUV these specimens did not do so. In view of the mauve-brown colour there cannot have been many examples of this material on the market.

Some variations of colour in single-crystal quartz have become surprisingly prominent over the past twenty years. The best known is known in the trade as 'ametrine'; it is not too difficult to work out that the name is a combination of 'amethyst' and 'citrine'. While most natural ametrine is found in Bolivia the material is popular enough for a synthetic version to have been grown by the hydrothermal method, fashioned and sold.

A paper in Gems & Gemology, Summer 2004, confirms that there is still some difficulty in distinguishing between natural and synthetic amethyst. Most commercial synthetic amethyst is grown in alkaline solutions on seeds and at growth rates that prevent the capture of the citrine-forming impurity. A band at 3,543 cm^{-1} is found in the IR spectra of almost all synthetic amethyst. However, the orientation of the seed may, in some circumstances, prevent the formation of this band and so detection by means of the IR spectrum may be impaired.

Nor is the 3,542 cm^{-1} band as rare in natural amethyst as was at one time believed. Confirmation of an amethyst's origin needs to rest upon the presence or absence of this absorption together with an examination of internal growth structures (Dauphiné and Brazil twinning, the presence of particular sectors and zones and a stream-like structure seen in the synthetic material). Natural amethyst's inclusions (chlorides, barite, calcite and multiphase fluid inclusions) are also helpful.

COLOUR ENHANCEMENT

The agate deposits around the town of Idar and Oberstein in Germany were worked over several centuries and at least by the nineteenth century the possibilities of dyeing were well understood; they have been summarized by Nassau in Gem Enhancement 1994. Heat treatment of carnelian to deepen the colour was begun in 1813 and the sugar-acid process for producing black agate was acquired from Italy in 1819. Yellow produced by the use of hydrochloric acid was known by 1822 and improved later; a similar mastery of the use of ferrocyanide to obtain blue was in place by 1845.

Dreher in Das Farben des Achates published in 1913, gives one of the first summaries of processes that were for long kept as a secret of the trade. Nassau makes the point that, although not all the secrets of these processes were disclosed, it is known that aniline dyes were avoided as they are prone to fading. Instead, colouring agents, and iron in particular, were used. This could give a strong red (iron nails were dissolved under heating in concentrated nitric acid). Several stages of soaking and washing were needed. Green was obtained by the use of chromic oxide, nickel also being recorded as being used on some occasions. Black onyx in nature shows a fine deep black, often with a white stripe and the effect was produced artificially by dissolving sugar in warm water. After an immersion of two to three weeks the agates, still wet, were transferred to concentrated sulphuric acid. The acid removes water from sugar, leaving the black carbon behind. A similar process may be used to darken the background of opals.

Blue could be obtained, according to Dreher, by two different processes. One used a solution of the very toxic potassium ferricyanide, in which the agates were soaked then transferred to a solution of warm ferrous sulphate containing a few drops each of sulphuric and nitric acids per litre. The other used potassium ferricyanide plus an iron salt.

Readers should certainly not try any of these very dangerous methods, unless working in a properly furnished and equipped laboratory, under trained supervision.

Agate, with its porous nature, and some other varieties of chalcedony seem ideally suited to these forms of treatment but it has to be said that the colours can quite often approach the garish, and most dyed specimens are quite obvious. As in some jade (jadeite) specimens, bleaching can be used to remove brownish or yellowish iron staining from pale chalcedony but it seems hardly worth the effort unless the stone is exceptional.

The chalcedonies are not synthesized, as natural material is plentiful. It is not only the agates that are dyed, however; the popular ornamental material known as tiger's eye is sometimes too dark a golden-yellow to brown for popular taste and the colour is lightened. Before examining the process we should note that tiger's eye is known to mineralogists as a pseudomorph. This can be defined as the replacement of one material by another, in the process of which the replacing substance takes the form of the one replaced. This can take place between two different minerals but in many cases inorganic mineral material replaces organic ones.

Tiger's eye is the replacement by silica of an original mineral with asbestiform (fibrous) characteristics which can still be seen in the 'new' ornamental material. For those who use tiger's eye with its very attractive stripes (which once were fibres) as worry-beads etc., there is no danger as the asbestos mineral has quite gone. Opal can very attractively replace plants or fossils in the same way.

When tiger's eye is too dark the colour may be lightened by removing some of the iron in the tubes left by the decomposition of the asbestos with a saturated solution of oxalic acid. Chlorine-containing bleaches may also lighten the stone. Iron in the tubes can be removed with hydrochloric acid, leaving the stones a greyish colour.

The hollow or partly hollow tubes may be filled with dye but it is not clear how widespread the practice may be.

Aqua aura (crystals and occasional cut stones) is gold-coated quartz.

CHAPTER TEN

The Jades

PROPERTIES

There are different minerals called jade, sufficiently distinct from one another for gemmological testing to distinguish them – although the unaided eye can be deceived, especially in dark green artefacts. They are both often found as carvings, rarely if ever faceted but common as cabochons. Both are revered, especially by the Chinese, not merely for their appearance and feel but because they are hard enough to allow characters to be engraved upon them. In this way many jade artefacts carry important poems, aphorisms of emperors and other valuable materials to be preserved indefinitely.

One of the jades is nephrite, a silicate of calcium, magnesium and iron, which at present is classed as a variety of the mineral actinolite, itself a member of the amphibole group of silicates. While a mineral may begin life in the literature as an individual species, increasingly sophisticated methods of chemical and structural analysis have encouraged the allocation of minerals to groups, some of them containing many individual species. The important amphibole group of silicates was revised in 1997 and contains over eighty species. So although nephrite was a distinct species, it is now part of a large group. But this does not really matter if all you need to know is whether your specimen is nephrite or something quite different.

Jadeite, a silicate of sodium, iron and aluminium, is a member of the pyroxene group of silicates. Unlike nephrite, however, it retains its individual species status within the group.

This chemical background is important as in the case of jadeite it allows the gemmologist or research scientist to identify specimens by their absorption spectrum and to recognize the particular structure of both minerals.

While neither nephrite nor jadeite is among the hardest of ornamental minerals both are tough. This is due to their structure of minute interlocking crystallites or fibres, rather than one single crystal per specimen. This structure allows dyestuffs to penetrate and settle, and it is the alteration of colour in jadeite in particular that earns jade a place in a book dealing with artificial and treated gemstones.

Most jade is green. Archaic Chinese artefacts are nephrite; jadeite is a rather brighter green than nephrite, and is also found in yellow, lavender and some white colours. Nephrite is characteristically a spinach-green but a whitish variety known as 'mutton-fat' is also typical.

The name 'imperial jade' is now proscribed by the trade in general and by the auction houses in particular. It was used to signify emerald-green jadeite with no speck of any other colour, especially white. Imperial jade has also to be notably translucent. It commands very high prices and quite a small cabochon may cost many thousands of dollars. It is not surprising, therefore that the name, which is a magnificent trade-mark, persists out of the public eye and is used for materials that are not either of the jade minerals, nor that it is widely imitated.

ENHANCEMENT AND IMITATION

There is no commercial synthetic jade but the search for an ideal imitation of the finer qualities of both types continues. Work on imitations has concentrated on the improvement of genuine jade and in some instances this is highly successful and the results hard to test. Work in this area is on jadeite rather than nephrite and in recent years some very ingenious imitations have reached the jade markets.

In 1998 I asked stall-holders at the celebrated Hong Kong jade market whether they had a particularly successful jadeite imitation, but they did not. One was found from elsewhere later, however. The Gemmological Institute of Hong Kong has also produced a three-box set of jade specimens, including their commonest natural imitations and examples of treatments.

Enhancement

The main thrust of experiments on jade treatments centres on jadeite. Natural jadeite which does not need enhancement is called A-jade. Jadeite containing unsightly dark brown or yellow patches of iron-staining which have to be bleached away by chemical action is B-jade. C-jade is not bleached but is impregnated to give a better colour. Jadeite which is bleached and impregnated is B+C-jade.

If a drop of concentrated hydrochloric acid is placed on the cleaned surface of a jade specimen it will be drawn by capillary action beneath the surface in a Type A specimen and an aureole will persist for some minutes at the point of application. With a Type B specimen the acid will stay on the surface before it eventually evaporates because the surface has been sealed by impregnation with the polymer. The acid droplet is magnified on the surface of Type C jade.

Testing has to begin with the nature of the specimen to be examined. Whatever the state of the untreated specimen or its final appearance, it will be possible to establish, with normal gemmological tests, its mineralogical identity.

Jadeite has an RI of 1.654–1.667 (the two edges may be clearly seen, even though jadeite can be classed as a crystalline aggregate) and an SG of 3.33. Its membership of the pyroxene group of silicates is shown by the presence of a strong absorption band at 437 nm (the pyroxene band). Absorption features arising from the colouring elements can also be seen: chromium, responsible for the fine green colour, shows its presence with absorption features reminiscent of those seen in emerald, although the absorption bands and lines are not quite so clear.

Green jadeite, unlike emerald (translucent varieties of which are the closest approximation to imperial jade) remains green when viewed through the Chelsea colour filter – emerald of the quality which could simulate the best green jadeite shows red.

The surface of jadeite is marked by dimpling, which gives a pleasing feel greatly appreciated by connoisseurs. The polishing of jadeite by diamond powder somewhat reduces the dimpling effect and may slightly detract from its value. Lack of dimpling might just persuade the gemmologist that a piece might well have been treated, by reason of its presumed date of polishing, but this would be quite a long shot.

In the Winter 2002 issue of *Gems & Gemology*, a piece was described. It was a variegated green bangle bracelet which was proved by standard gemmological tests to have been dyed.

It gave an RI of 1.66 and the pyroxene absorption band at 437 nm. It also had the absorption band at 650 nm which is indicative of dyeing. Like most bands associated with dyestuffs, this band is rather woolly than sharp.

Under magnification colour concentrations could be seen along grain boundaries. Under reflected light and with magnification an unusual surface texture could be seen, which was ascribed to the preferential erosion of some grains during the bleaching process. Randomly interlocking jadeite grains become visible as the acid-soluble minerals between them are dissolved. This process makes narrow outlines round the grains and cavities between them, which are filled with a polymeric material or a wax.

By reflected light and under magnification differences in lustre between the jadeite grains and the filled cavities were apparent. The piece fluoresced a medium mottled greenish-yellow under LWUV and a very weak mottled yellow under SWUV. These effects are also characteristic of impregnated jadeite. Final confirmation of impregnation was provided by IR spectroscopy. When the bangle was viewed in combined transmitted and reflected light with reinforcement from fibre-optic sources some dark inclusions could be seen in an area with lower lustre. These looked like the minute green spherules and gas bubbles seen in the polymer filling of a bracelet previously examined. In this example the inclusions were a dark violet-blue (although the bracelet was dyed green) and could be seen only in a small part of the filler.

I have highlighted this particular example as it shows the kind of work now facing those who routinely examine the jade minerals.

The colour enhancement of jade may begin with the heating of specimens whose colour is considered unacceptably dark for commercial success. Jadeite is not irradiated as a general rule. Nephrite is not commonly dyed.

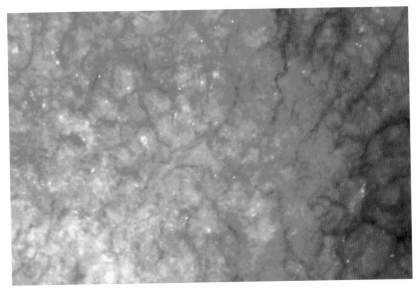

Dyed jadeite. The green colour will not deceive anyone familiar with jade but how many can claim this expertise?

If iron compounds showing yellow or brown are present in jadeite, heating may turn it brownish or reddish. Lavender jadeite is notably sensitive to heat and according to Nassau in *Gemstone Enhancement*, temperatures of only 220–400°C may be sufficient to bleach it.

Some dyes may fade over time in ordinary daylight, aniline dyes being more prone to this than others. Colour concentration should be the first sign sought by the gemmologist, in just the same way that dyed quartz varieties are examined. In carvings the best place to allow colour to concentrate is somewhere at the back of the specimen so this area and the base should be examined carefully with the lens.

Resins used for impregnation can be identified by infra-red spectroscopy, a characteristic band occurring at 2,900 cm^{-1}. GIA has reported a lavender cabochon of 15.86 ct which was identified as jadeite. Small cavities containing transparent colourless filling material were present – IR spectroscopy showed this to be a synthetic resin.

Sometimes the surface of jade can be improved by varnishing; a specimen was reported by GIA in 1994, a large greyish-purple bead coated with a layer of mottled green varnish. Some of the varnish had spalled off, showing the original colour beneath. The coating gave an RI of 1.52 compared with the 1.66 of jadeite and the SG was 3.29 compared with 3.33.

When a thermal reaction tester was applied the surface coating melted and the specimen gave a strong chalky-blue fluorescence under LWUV with a weaker response to SWUV. Fluorescence is not a feature of jadeite, although some specimens may show an uneven yellowish-white response.

In another coated specimen, a pipe, tested by GIA two layers had been applied. When the upper coating was chipped the layer beneath was revealed. The lower layer was a mottled green and the upper one colourless or a uniform very light yellow. The two layers could be seen when the specimen was examined in UV. Both layers showed a yellow-green with varying intensities. No fluorescence was observed from the jadeite, whose identity was established from the presence of the pyroxene absorption band at 437 nm.

IR spectrometry showed that the two coatings were organic poly-mers. With a single IR test and nothing more the specimen could have been mistaken for B-jade.

Jade treated by bleaching and impregnation may show what has been called a 'beehive' effect. This comes about when bleaching is used to remove iron staining, leaving a honeycomb structure formed by grain boundaries. In a good-quality green and white patterned bangle tested by GIA the voids left by the removal of the iron impurities were filled with a colourless transparent polymer or wax; grain boundaries could be seen through the microscope.

While jadeite is by far the more likely of the two jade minerals to be treated there have been reported cases of dyed nephrite. The green dye can be detected by the characteristic dyestuff absorption spectrum giving a band centred at 660 nm.

The presence of iron staining in jade should not always call for bleaching. In nephrite, for example, iron staining is often the result of long burial and should certainly not be interfered with on archaeological grounds. In some cases a specimen may be treated in such a way that some of the iron stains are left behind to suggest authenticity. In some other cases the acids used in some forms of treatment have been found to leave behind yellow exudations, which of course give the game away.

Waxing is most commonly used to conceal small fractures close to the surface. Paraffin wax has often been used: it is easy to obtain and can be introduced into surface-reaching flaws without difficulty. Interestingly, Nassau in *Gem Enhancement* made the point that this type of enhancement may not be noticed when the specimen comes to be set in jewellery and the filled flaw could be damaged. In general transparency may be improved by some forms of impregnation and coating.

Imitation and Synthesis

A synthetic jade (jadeite), for example, was reported in 1984 from the Inorganic Materials Laboratory of the General Electric Company. No examples of the product were sold. It was believed that synthesis took place from a starting material of melted sodium carbonate, alumina and silica, forming a glass. Growth was reported to have taken place in platinum crucibles at temperatures around 1,500°C. Chromium would need to be added if a green colour in any way resembling that of jadeite was to be achieved.

No details of the properties seen under magnification have come to my notice and I should not think that the product would have more than an experimental significance (though that is considerable), as the jade minerals are not excessively rare and convincing natural simulants are plentiful.

Jade may occasionally be synthesized for experimental purposes. It is, however, inevitable that valuable and interesting minerals will be imitated. There are plenty of imitations around, many being quite hard to test if only because their existence is unsuspected.

The main dangers could be said to come from other natural materials which can resemble either of the jades quite closely. These, however, are outside the scope of this book – they are dealt with in O'Donoghue and Joyner: *The Identification of Gemstones*. Imitations made from composites, however, *are* within the scope of the present work, and in some ways they need all the publicity they can get.

One very dangerous example is formed from jadeite itself. A cabochon of jadeite is hollowed out in such a way that the upper surface of the piece removed can be dyed a fine green colour; the same dye can be applied to the inner surface of the other portion. The two sections are reunited in such a way that the join can only be seen by the most careful examination under magnification.

Clearly the colour must be fine enough to deceive or there would be little point in taking such time and trouble to assemble such a composite. My feeling is that they are not very common or there would be more unfavourable publicity about them. Identification begins with suspicion (both the colour and the translucency of fine jadeite are seen) and continues with the use of a spectroscope, which will show the chromium spectrum and presumably the pyroxene band at 437 nm, indicating that the specimen is jadeite. However, a careful examination of the absorption spectrum will show the woolly, indistinct absorption band in the red which seems always to indicate the presence of a dyestuff.

It has been reported from time to time that some manufacturers of such composites have left ridges where the join comes at the base of the cabochon, but this should definitely not be taken as an indication of the true nature of the specimen, one way or the other.

Interesting examples of man-made simulants can be found in the literature. In the Summer 1996 issue of *Gems & Gemology* a composite imitating a statuette is described. The specimen weighed 239.37 ct and was made from a plastic substance and calcite, dyed to resemble jadeite. The specimen was tested on a refractometer which disclosed the strong birefringence of the calcite (note that testing large pieces on a refractometer can pose problems); the SG was 1.98 (compared with the 3.33 of jadeite).

Under magnification the statuette was found to show white grains in a groundmass of transparent colourless or green material. The lustre was resinous but GIA concluded that the imitation was quite successful.

In the same year GIA described a cabochon which appeared to be a fine green when viewed from above but which when examined from the side showed a very thin green layer on the top with a thicker white layer beneath. The layers were cemented together with a yellowish adhesive containing gas bubbles. The green layer was about 0.1 mm thick and the white 2.2 mm. The dark layer was mottled with colourless veining and the white layer showed a distinct crystalline structure. RI readings from the two layers were 1.64 and 1.74. An absorption spectrum characteristic of chromium could be seen and X-ray diffraction showed that both the layers were jadeite.

In the early 1970s a material known variously as Victoria-stone or meta-jade was suggested as a jade simulant but the resemblance was not very close. It was a glass containing angular fibrous inclusions, translucent at best, in a mainly transparent body. The RI was 1.48, well away from the 1.66 of jadeite, and there was general absorption above 590 nm and below 510 nm. There was no fluorescence nor change of colour when specimens were examined through the Chelsea filter.

Another glass jadeite imitation was reported by GIA in 1995. The colour was apple green, inclining to yellow, and with circular whitish areas with deeper colour and transparent surrounds. Gas bubbles were profuse, some appearing as surface cavities and making identification as glass easy. The whitish areas were seen under magnification

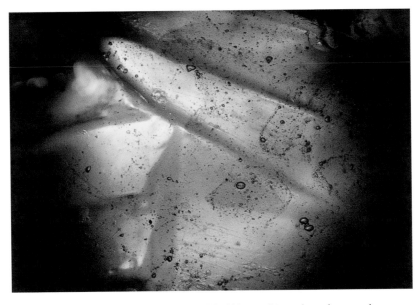

Paste (imitating jadeite). Large and small bubbles and irregular colour patches are seen in this green glass

to be associated with bundles of fibrous inclusions. Devitrification (the separating out of some of the solid materials from which the glass was formed) was probably responsible. This glass imitation, seen in Vietnam, was found to contain rubidium, yttrium and zirconium, elements not present in jadeite.

Another mineral treated to resemble jade was fashioned into a mottled lavender, green and orange bangle. It gave a refractive index of 1.54, which immediately suggested quartz. Dye could be seen concentrated in fractures and between grains. With a hand spectroscope a dark absorption area could be seen in the deep red but there were no other elements. A medium pink fluorescence could be seen from the lavender areas under LWUV, the whole piece giving a medium to strong bluish-white under SWUV. Advanced testing showed that the piece was not jadeite but a dyed quartzite with polymer impregnation. It is not common for quartzite to be impregnated.

Two large glass imitation jade carved figurines were reported by GIA in the Winter 2002 issue of *Gems & Gemology*. The pieces, from 30 to 50 cm in height and a translucent to opaque very light greyish-green, were reported to have come from Asia and to have been sold as jadeite, with notably elaborate and careful packing (such details are worth bearing in mind when items are sent from another part of the world).

Both pieces appeared to have been fashioned from the same material. The polish was reported as poor and both RIs were in the low 1.50 area (jadeite is 1.66). Both fluoresced a very weak yellow under LWUV and weak to medium yellow under SWUV. No crystalline structure could be detected on close examination: a few chipped areas showed conchoidal fracture with a vitreous lustre.

Using fibre-optic illumination small round and elongated gas bubbles could be seen by the unaided eye and swirls characteristic of glass were also visible. A test in an inconspicuous area showed that the hardness was greater than 5 so that plastic could be eliminated. The final identification was that the material was manufactured glass.

GIA makes the point (as I have often done and still do to students) that 'traditional' synthetics and imitations do not just fade away when some new material or technique makes its appearance. Glass is

so easy to come by and to work; in this case under review the elaborate packing and the cost of postage may have come close to the
actual cost of the carving of the figurines.

CHAPTER ELEVEN

Topaz

Topaz is a fluosilicate of aluminium with the formula $Al_2(F,OH)_2SiO_4$. Crystals are often fine, large and clear, with a hardness of 8, an SG of 3.53, an RI in the range 1.62–1.64 and a birefringence of 0.008. These properties vary somewhat with colour but not enough to cause confusion.

SYNTHESIS

The easy and perfect cleavage of topaz may cause problems for the lapidary, and is probably one of the many reasons why topaz is not generally synthesized. Another is that while the process is possible, the resulting crystals would be far too small for ornamental use. Generally speaking the synthesis of silicates is quite difficult – oxides are much easier – and in the case of topaz the natural material is sufficiently plentiful. Citrine makes more than an effective substitute for the yellow through orange to brown topaz.

In *Gems & Gemology*, Winter 2001, however, a hydrothermally grown topaz was reported from the Institute of Experimental Mineralogy, Chernogolovka, Russia. One product was described as a semi-transparent light grey crystal of 23.66 ct and another, a semi-translucent brown crystal, weighed 27.48 ct. Two-phase inclusions were present: other gemmological properties were in the range of the natural material. Since all topaz, to be of commercial use, needs to be completely transparent the Russian product seems not to be a threat.

ENHANCEMENT

Blue topaz is not too hard to obtain but the colour is rather pale, and comparison with aquamarine is inevitable. In recent years a much stronger and darker, sometimes rather metallic blue topaz has been offered by jewellers and examples can be seen on display almost anywhere. Prices are not unreasonable; the problem is the usual one of identification.

Blue irradiated topaz shows neither internal nor external signs of treatment – only the deep blue gives it away. Most natural blue topaz is paler

With aquamarine we have found that greenish-blue material has been treated to give a stronger blue for so long that no one ever thinks the practice needs disclosure or explanation. Nor have special names been coined for the heated product. This treatment of pale blue topaz to give a stronger blue has been accepted very quickly, although one or two names have been invented in this instance.

Colour in the yellow to brown and the blue natural topaz is the result of colour centre operation. Save in a few instances, in normal conditions these colours are stable. Some yellow-to-brown natural topaz, notably some examples from Utah, will fade quite quickly in conditions of strong light; specimens showing fine crystal forms are kept in boxes tailored to their shape in the collections of the Natural History Museum in London so that their colour is not lost. It is worth pointing out that these unstable-colour stones are in fact rare and that the colour of almost all yellow to brown topaz is stable.

However, some colourless topaz can be irradiated to give a brown colour, with some blue (more brown than blue). Heating will drive off the brown and leave the blue intact; this blue is stable and on sale widely today.

It is just as well that disclosure does not seem to be called for, because the treated blue topaz cannot be distinguished by normal gemmological tests, save by colour, from untreated blue material. One popular name for a rather inky dark blue is London blue, and there are also American blue, Super blue, Swiss blue and no doubt others.

The process of irradiation normally takes place in a linear accelerator (linac) and in early experiments some specimens became radioactive, but no blue topaz on the market today should give any cause for alarm. Scares of this kind, though usually based on one or two actual occurrences, are not common. As a matter of interest, the London laboratory of GAGTL was sent a steel-blue topaz with twenty counts per second on a Geiger counter – this amount of radioactivity would make such a stone unsafe to wear.

After irradiation stones are heated as mentioned above, although if they have been irradiated in a nuclear reactor subsequent heating does not always seem to be necessary. A tendency to inkiness can be reduced by heating, however.

In the unlikely event that it is necessary to distinguish an irradiated topaz from a natural one the presence of radionuclides can be detected by gamma-ray spectroscopy.

A green topaz, with the name ocean green, obtains its colour from irradiation but specimens do not seem to have been widely sold and in any case the green is strongly inclined to yellow and not very pronounced.

Pink topaz is found in nature; the colour arises from the irradiation of material which was originally brownish-yellow, and it is not possible to determine whether or not in a particular specimen the colour arose naturally or artificially. Fortunately it does not matter in the commercial context (although the pink stones are very attractive).

Pink topaz owes its colour to chromium and shows an emission line at 682 nm.

A topaz with an unusual orange colour was shown to the German Gemmological Association for identification. It was faceted in an oval shape and weighed 2.25 ct. The colour resembled the orange of the Namibian spessartines, and for that reason was offered for sale as 'mandarin topaz'. When examined with the unaided eye violet reflections could be seen and when the stones were immersed they resembled a doublet. The crown was seen to be colourless and the pavilion an orange-brown with no cementing layer visible. The stone was not a composite but a colourless topaz which had been coated on the pavilion.

The Winter 2002 issue of *Gems & Gemology* noted an orange topaz coated with hematite. Two bright orange stones were examined by the SSEF Swiss Gemmological Institute, which received them under the understanding that they were topaz. One weighed 6.97 ct and the other 2.95 ct. Both were reported to have come from Nigeria, where they had been offered for sale. Confirmation of topaz was provided by the RI of 1.610–1.619 with a birefringence of 0.009 and SG of 3.53. Both were inert under both types of UV. The colour was unusual for topaz.

Under the microscope both stones showed a bluish-green iridescence on the back facets. Small colourless chips could be seen along the back facet edges when bright-field illumination was used and the back facets also showed some colourless scratches. All of these were indicative of coating and there have been reports of red, orange and pink coating of topaz in the past, a sputtering process being suggested.

A previous report said that such a coating can easily be scratched. In this case, however, a needle point failed to remove the coating, something which had not previously been described with coated topaz. Moreover, while some coated stones show a spotty appearance which is considered characteristic of coating, the topaz specimens in this case did not show this effect.

The pavilion surfaces were examined using EDXRF chemical analysis, which showed high concentrations of iron. Raman spectroscopy on the same surfaces gave distinct peaks for hematite as well as peaks showing that the material beneath the coating was topaz. The conclusion was that the orange colour was caused by a thin coating of microcrystalline hematite applied to the pavilion surfaces. There appear to have been no previous reports of synthetic hematite being applied to the surface of a gem material. The note in *Gems & Gemology* makes the point that all unknown faceted specimens should be inspected from all angles, not just through the table facet. Some stones coated in this way have been offered as diffusion treated but no diffusion took place in the case described.

CHAPTER TWELVE

Turquoise

Turquoise would cause considerable trouble to gem-testing laboratories if it were valued more highly, but it cannot compete in price with fine rubies or emeralds. It does have its own following, however, sufficient to make synthesis profitable. Like lapis lazuli, it is fairly easy to imitate and it also lends itself to colour enhancement.

Turquoise is a phosphate of copper, which provides the colour. The blue shades into green, which has been preferred by some connoisseurs. The material is porous and thus easy to dye, the aim being to darken the colour. Waxing the surface is the method most commonly used, although it is easily detected when the thermal reaction tester is brought close to the surface.

SYNTHESIS AND IMITATION

In the early 1970s Pierre Gilson, better known for his synthetic emeralds, produced a synthetic turquoise or turquoise imitation, which was marketed as Gilson created turquoise. The production technique was supposedly similar to that used in the manufacture of ceramics (Gilson's previous work was in the field of refractory materials). Presumably the turquoise manufacture began with the production and consolidation of a powder. Nassau, (1994) discounts the possibility of ground-up natural turquoise being used as it contains some iron, which is not found in the artificial product. Gilson also makes a 'veined' turquoise which could be deceptive, as much natural turquoise is veined.

The Gilson material has properties close to those of natural turquoise, whose typical values are: SG 2.75–2.81, and RI 1.61–1.62. Gilson turquoise's values are 2.72 and 1.604 respectively. Both it and natural turquoise have a hardness about 6. Any stone offered as turquoise and showing an SG lower than 2.65 should be tested further. In practice, testing for SG is rarely done outside gemmological laboratories.

The copper absorption spectrum, showing two rather faint bands in the blue at 460 and 432 nm should distinguish turquoise from similar stones; they have rarely been noted in the Gilson product.

Examined under magnification in the range 30–40x the Gilson stones showed a characteristic surface marking of angular dark blue particles against a whitish ground-mass. This was not seen in a sample of seventy specimens of natural turquoise in a test reported in 1973. No response was noted under UV.

Gilson used the name Cleopatra for a medium-blue product and Farah for a dark blue version. As well as veined and unveined specimens, Gilson turquoise may consist of turquoise itself with one or two additional phases or it may be a substitute made almost entirely from calcite. This material would, of course, effervesce with the acids sometimes used in gem testing. The Gilson material is best summarized in *Gems & Gemology*, 15, 1979.

As well as the Gilson material other artificial simulants of turquoise have been produced. The name Turquite was given to an aluminium-poor material produced by Turquite Minerals of New Mexico, USA. Though poor in aluminium, Turquite is rich in sulphur, calcium and silicon.

An imitation of turquoise has also been made by the Syntho Gem Company of California, and has properties similar to those of the Gilson material. The Adco Products turquoise, also manufactured in California, also resembles the Gilson turquoise. Of these imitations only the Gilson product has the same crystal structure as natural turquoise.

Imitations of turquoise are far more likely to turn up in jewellery than synthetic products. Imitations may be tested by placing a drop of Thoulet's solution on the surface. As this solution is an iodide of potassium and mercury, health and safety regulations are likely to apply to its use, and it would be better to rely on experience acquired with types of magnification. If Thoulet's solution is used the surface of the specimen will turn brown if it is an imitation; true turquoise remains unaffected.

Viennese turquoise and Neolith are turquoise imitations of different compositions. The name Viennese turquoise has been given to a precipitate of aluminium phosphate which is pressed for consolidation and coloured blue by copper oleate. Neolith is a mixture of copper phosphate and the aluminium hydroxide mineral bayerite. In both cases the surface of a specimen will show yellow when touched by a drop of dilute hydrochloric acid.

Gemmology Queensland, June 2003, reported briefly that an imitation of turquoise had been produced in Germany. High pressures and temperatures are said to be involved and the resulting material contains no epoxy resin though 'a special bonding agent is used'. The report came from Sanwa Pearl Trading.

Stained magnesite (SG near 3.00) and stained howlite, SG 2.53–2.59, have often been offered as turquoise. Magnesite will effervesce with warm dilute hydrochloric acid.

ENHANCEMENT

In *Gemstone Enhancement*, Nassau reminded readers that the preferred colour of turquoise varies between races and countries. Buyers in Egypt are said to prefer a greenish to a bluish colour and blue material is altered to green.

Turquoise can be coated, both to improve the general appearance of the stone but also, in some instances, to assist in consolidation – much turquoise tends to be pale and powdery, and any sort of treatment will darken it to some extent. Nassau described a string of turquoise beads whose surface was painted blue with some black to provide an imitation of natural matrix. The beads were coated with a clear lacquer after this treatment.

Turquoise is one of the most difficult gem species to identify with certainty. A microscope, already described as vital for the testing of the Gilson productions, is also essential if the gemmologist wishes to establish whether or not a particular specimen is coated. Examination of the drill-hole will sometimes show a junction between a bead nucleus and a coating.

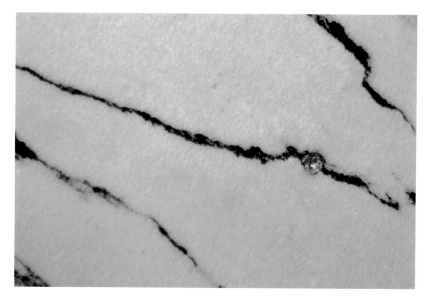

Neolith imitation of turquoise. A drop of hydrochloric acid on the surface will turn yellow

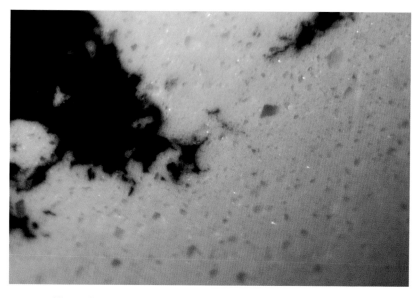

The surface of Gilson imitation turquoise shows dark bluish spots in a whitish groundmass

Using a thermal reaction tester (hotpoint) may cause some surfaces to show local melting and if the specimen has been coated with plastic the characteristic pungent smell will develop.

While ammonia will remove some of the dyes used to alter the colour, this is not something that would be done often. In general turquoise that has been treated will show a higher content of potassium, but here again most gemmologists and dealers would not normally take turquoise testing so far. And a strong copper absorption from a specimen offered as turquoise suggests that some form of treatment has been carried out.

Nassau explained that the body of turquoise varies from compact to porous and that very porous material may not show blue on account of light scattering from the pores. When the pores are filled with an RI higher than that of air (e.g. water, oil, wax or plastic) the blue colour can be seen again. When the colour of a particular specimen is said to have faded it is commonly due to the drying out of one of these substances.

Turquoise colour may also be affected by skin oils and some cosmetics; the usual result is a change from blue to green, the degree of change depending upon the porosity of the particular specimen.

CHAPTER THIRTEEN

Lapis Lazuli

apis lazuli is a rock rather than an individual mineral. It is composed of a number of minerals of which the most important are lazurite (which gives the colour), haüyne (also a bright blue), sodalite (a darker blue) and nosean. Calcite, diopside and pyrite are usually found; a high content of the former lowers the value considerably. On the other hand the bright brassy yellow flecks of pyrite enhance the general appearance.

Since it is a rock the properties vary; the RI (which can be taken only on flat-backed cabochons – much lapis is carved) will be near 1.50 and the SG 2.7–2.9 (sometimes higher if there is a high pyrite content). Some may show orange fluorescent streaks in LWUV light. This is commonest in Chilean lapis, which contains a good deal of calcite and some specimens may show pink under SWUV light. A drop of hydrochloric acid placed on a specimen may produce a smell of rotten eggs but I would hope that the acid test is used infrequently; if the acid is placed on the calcite part it will cause effervescence.

IMITATION

Because it is a rock, it might be assumed that it is hard to imitate lapis successfully but specimens are large enough for fine blue cabochons and small carvings to be obtained without using any material which is not of the best colour. So all the imitator has to do is to produce the colour and this is very well done by a variety of synthetic spinel said to have been placed on the gem markets in 1954

(I mention this in case any readers know the date of manufacture of their lapis jewellery).

The material is grown by flame fusion, doped with cobalt and fused at a temperature of not less than 2,000°C. The name 'synthetic sintered spinel' was used for a while but does not turn up now.

The spinel's SG, at 3.64 is much higher than the 2.7–2.9 of lapis but few will want to do SG tests today when other methods are available. It is easier to examine a suspect under a strong light with a Chelsea filter. As cobalt is the cause of the colour it will cause the specimen to appear a strong red. Some lapis specimens may show a dull brownish-red with this test but there should be no confusion. In some instances small flecks of gold are added to simulate pyrite ('fool's pyrite') but no attempt is made to imitate the calcite for obvious reasons. Many examples of the spinel will have flat backs (in cuff-links, for example) and they will enable an RI reading to be obtained – it will be around 1.728 compared with the 1.50 mean figure of lapis.

The spectroscope will show cobalt bands in the red, yellow, green and blue regions where lapis lazuli shows no absorption in the visible. If a thin enough piece of the spinel imitation can be obtained, it will show a distinctive reddish-purple colour.

Stained dolomite (magnesium calcium carbonate) has also been reported as an imitation of lapis lazuli. It will react with warm acids (dilute hydrochloride acid) by effervescing. It has an RI of 1.50–1.68 and an SG of 2.85 (near lapis lazuli); it would not be easy to carve as it cleaves so easily and signs of this would be apparent.

Gilson Lapis Lazuli

Pierre Gilson brought out an imitation of lapis lazuli in the 1970s. Several varieties were offered, some containing added pyrite. The material was found to be porous and for that reason has a lower SG than natural lapis lazuli, at about 2.46. When a piece was drawn across an unglazed porcelain streak plate it left behind a strongly coloured blue powder; true lapis lazuli leaves a much weaker blue. When this test was carried out on the Gilson material a distinct sulphurous smell was noticed; a similar smell but much less distinct occurs with the natural material but in this case a greater pressure needs to be applied to the specimen and plate.

If tested with acids the Gilson material reacts more strongly than natural lapis lazuli. The composition is ultramarine (an intense blue pigment) with hydrous zinc phosphates.

ENHANCEMENT

Natural lapis lazuli is frequently dyed or waxed to improve the colour and lessen the effect of any calcite. Specimens should be examined under magnification to see whether or not dyestuffs have concentrated in cracks. Dye can often be removed with an acetone-soaked swab (if prevailing health rules allow it); nail polish remover will do as well.

Natural material may occasionally be dyed to give what is thought to be the lapis lazuli colour. One of the commonest examples (though it is not really very common) is howlite, which has an SG of 2.45–2.58, well below the 2.83 of lapis. Unstained howlite from California gives a brownish-yellow or intense orange fluorescence under UV light and a specimen dyed to resemble lapis lazuli showed the same fluorescence colour in a part of the stone from which the dye had been removed.

Just as soil in the garden appears darker after rain, so a light-coloured specimen of lapis can be darkened by impregnation with plastic material. Chilean material is usually chosen for treatment since it diminishes or removes the whitish calcite areas. A thermal reaction tester will produce a pungent 'plastic' smell.

NOTES FROM THE LITERATURE

Plastic imitations of lapis lazuli (and malachite) have been used as inlays in watch-face material. Watches with natural lapis faces will feel much heavier. Unpolished edges between each link of the inlay will appear uneven and grainy where plastic will be smooth.

An imitation of lapis lazuli was reported whose colour could not be removed by acetone even though the beads of the necklace were reported to have stained skin and clothing. In the end denatured alcohol (found in most scents and colognes) did the trick.

In 1986 GIA described a lapis lazuli necklace whose beads were violet-blue and very deeply coloured. They fluoresced a patchy red under LWUV light while under SWUV light only a few gave the usual chalky green response of natural lapis lazuli. An acetone swab did not remove as much of the colour as expected but paraffin treatment was confirmed by the specimens sweating when a thermal reaction tester was brought near. Some of the beads showed a purple dye in cracks, visible under magnification, and the necklace showed a definite brownish-red through a Chelsea filter. This was a brighter colour than had been reported previously for natural lapis lazuli. It is likely that the beads had been paraffin treated and then dyed after the seal created by the paraffin treatment had been removed. The dye was strong enough to cause virtually all the colour.

In the Spring 1991 issue of *Gems & Gemology*, GIA reported an imitation of lapis lazuli which closely resembled the natural material, showing an even, dark violet colour and with randomly distributed pyrite grains. Gemmological testing gave an SG of 2.31 compared with the 2.83 of natural lapis lazuli and an RI of 1.55. The specimen did not react to LWUV but gave a weak chalky-yellow fluorescence under SWUV. The pyrite inclusions were proud of the surface, showing that they were harder than their host. Also visible were some random, shallow, whitish areas. The specimen transmitted more light from a fibre-optic guide than would be expected from natural lapis lazuli. The specimen became virtually invisible when viewed through a Chelsea filter using the same source of light. When the filter was used with reflected light the piece showed a slightly dark reddish-brown colour. A thermal reaction tester produced a weak acrid smell with a slight melting and whitish discolourations. It is possible that some kind of plastic binder may have been used; the specimen turned out to be barium sulphate with a polymer binding agent, which was proved by X-ray diffraction analysis.

A dyed blue calcite marble was described by GIA in the same issue as a possible lapis lazuli imitation. The item was a single-strand necklace with uniform beads which were believed to be lapis lazuli. The RI was 1.4–1.6, with birefringence which suggested a carbonate. When tested with a 10 per cent solution of dilute hydrochloric acid, effervescence could be seen. The specimen could not have been magnesite since magnesite will not react with such a solution at room temperature. An acetone-soaked swab removed the colour, and when the specimen was examined along the drill hole a yellow underlying colour could be seen. X-ray diffraction analysis proved the piece to be a dyed calcite marble.

Jewellery that stains the skin or clothes is not likely to command much in the way of sales. A necklace of opaque blue beads with yellow metal spacers tested by GIA and reported in 1989 were found to stain the skin, and dyed lapis was suspected. The RI was not that of lapis lazuli and not all the beads looked the same. There was no fluorescence under X-rays. Some of the beads were found to be dyed calcite while others were dyed jasper of the type once known as 'Swiss lapis'.

For a time a synthetic sodalite was grown in China and offered for sale as lapis lazuli; samples were heavily twinned and included. The specimens were colourless when grown, but later irradiated to give the blue colour.

Acrylic substances have been used to coat a number of different ornamental materials; wax and paraffin are the most common for the coating of lapis lazuli. In 1992 GIA quoted an article in which aerosol sprays were recommended for surface improvement. One of those recommended gave a transparent colourless surface which was tested and found to give a glassy coating on fashioned pieces of lapis lazuli and jadeite. Four light coatings were applied, the surface then showing a concentration of glassy material in irregularities and carved recesses. The coating was easily removed with a razor blade and also with acetone.

A glass imitation of lapis lazuli was noted by the ICA (International Coloured Stone Association) in 1991; a bead necklace and a single loose fashioned stone were tested. The material was said to be opaque and predominantly a medium blue with darker blue portions distributed in a marbled pattern. The RI was measured at 1.62 and there was no response to LWUV light, although a faint powdery blue could be seen under SWUV light. A uniform distribution of very small, highly reflective and transparent slightly brown flake-like spots could be seen under magnification, some with triangular outlines. These were presumed to be intentional imitations of pyrite crystals.

Another imitation of lapis lazuli took the form of a pair of scarabs. The colour was more like the dark blue of sodalite than that of lapis lazuli and it was found to be concentrated in fractures. X-ray diffraction showed the material to be a dyed feldspar.

In 1993 GIA reported 'howlite lapis' in the form of large violet cabochons. They were considered to be a quite convincing imitation of lapis lazuli as they contained white, dye-resistant veining quite like the calcite so often seen in natural lapis lazuli. When examined the materials were found not to be howlite (there was no strong bire-fringence) but the presence of a dye was confirmed by acetone-soaked swabs. Magnesite was proved by X-ray diffraction analysis. Some of the dye came off when the pieces were washed in a mild soap solution.

A lapis lazuli imitation for which X-ray diffraction analysis appeared to give the pattern characteristic of a phlogopite-rich ceramic was found when a thin section was examined to be predominantly a strongly birefringent mica-like material with high-order interference colours. There were also small singly refractive zones coloured dark blue which appeared black between crossed polars. Further testing with a scanning electron microscope with an energy-dispersive spec-trometer showed that the specimen was largely composed of crystals with a roughly rectangular outline and a lamellar structure, character-istic of mica. The spectrum showed that magnesium, aluminium, silicon and potassium were present as major elements, indicating that it was probably the mica phlogopite. Also present were grains of an

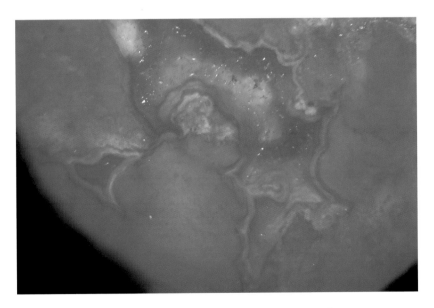

Dyed jasper may contain quartz flakes easily mistaken for pyrite

unidentified silicate of calcium and magnesium, perhaps diopside, as well as the mineral lazurite.

Attractive gold veining in a specimen of lapis lazuli proved to be cement with pyrite dust.

A glass imitation of lapis lazuli was made from a blue aventurescent material in which the spangles were triangular copper crystals.

The material long and unfortunately known as Swiss lapis is a dyed jasper and gives the optical and physical properties of quartz (SG 2.65, RI about 1.55).

CHAPTER FOURTEEN

Chrysoberyl

ALEXANDRITE AND VERNEUIL-GROWN CORUNDUM

The alexandrite and cat's eye varieties of the mineral chrysoberyl are two of the more valuable gemstones. There is also an 'alexandrite' which is a vanadium-doped synthetic Verneuil-grown corundum. Chrysoberyl is beryllium aluminium oxide, $BeAl_2O_4$, with a hardness over 8, RI of 1.74–1.75 and SG of 3.71–3.72. For purposes of comparison corundum has an RI of 1.76–1.77 and an SG of 3.99–4.00.

Even under low magnification the corundum 'alexandrite' can be seen to contain the characteristic curved growth lines of the Verneuil product (in fact, this is the best specimen for students to practise on). The considerable size of many of these 'alexandrites' should also warn the potential customer, as the majority of natural alexandrites hardly ever reach such sizes.

Alexandrite owes its position among the classic gemstones to its change of colour with change of lighting. The finest examples show a near-ruby to a dark raspberry red in incandescent (light bulb) light and dark near emerald green in daylight or under strip lighting. While some textbooks suggest that candlelight is also a good medium in which to see alexandrite's red colour, in practice many stones are dark and would appear only black unless the light was arranged to pass through them. The colour of the synthetic corundum imitation is nothing like that of natural alexandrite, appearing purple in incandescent light and a very characteristic slate-blue in daylight or under fluorescent lighting. Once seen, this material is not easily forgotten (particularly if some dispute has arisen over a specimen's identity).

Apart from the colour, the materials can be distinguished in other ways. A spectroscope will show a prominent absorption band in the blue at 475 nm and a microscope will show the prominent curved colour banding which is easier to see in this variety of Verneuil-grown corundum.

The difference in price between the synthetic corundum and fine-quality natural alexandrite is very large indeed, since it costs only few pence to make rough material of a size sufficient to provide large faceted stones. In comparison, the upper estimate for an alexandrite of high quality sold in the major salerooms over the past few years may easily reach many thousands of dollars.

SYNTHETIC AND IMITATION ALEXANDRITE

As we will see in the chapter on spinel there is an imitation alexandrite produced by chromium-doped spinel.

Alexandrite has also been grown by the flux-melt method, the original attempts going back as far as 1845 when Ebelmen, of synthetic emerald fame, grew chrysoberyl using borate fluxes. The first synthetic crystals large enough to be used commercially as gemstones were grown by Creative Crystals Inc., of San Ramon, California. The trade name used was Alexandria created alexandrite. Stones show flux inclusions of the type already described from ruby and emerald. The patent (US 3,912,521 of 14 October 1975) states that 0.7 per cent chromium was sufficient to give the alexandrite effect. The RI was 1.746–1.755 and the SG 3.73. These figures are comparable with the natural material. In some examples a microscope showed a layer of dust-like inclusions parallel to the seed face and colour banding was prominent. Smoke-like veils of flux were reported.

Alexandrite crystals have been grown by pulling but most if not all examples are grown for research and non-ornamental applications. In general, although chromium is present as a dopant, the crystals are not sufficiently coloured to make good gemstones.

A synthetic alexandrite which was probably grown either by crystal pulling or by the flux-melt method was reported in 1987. It was the Japanese Inamori alexandrite, which showed not only a colour change but a cat's eye effect as well. The colour change was quite pronounced and the eye moderately intense. Cabochons examined for the GIA study ranged from 1.04 to 3.31 ct. Specimens fluoresced a dark greyish-green with a faint purple overtone, with the eye giving a slightly greenish bluish-white. Stones in general appeared somewhat oily and dull. Examination with the 10x lens showed no distinctive inclusions, but it is interesting to note that when the stones were examined in the long direction under a strong incandescent light they showed asterism, with two of the rays noticeably weaker in strength to that of the eye. Such an effect has not so far been noted in natural alexandrite. Gemmological testing showed no distinction between the RI and SG of the Inamori material and those of natural alexandrite. However, when specimens were placed over a strong light they showed a strong transmission luminescence of a greenish-white colour. The same effect could be seen in strong sunlight and other types of artificial light. This is thought to be the cause of the oiliness.

Under SWUV the stones showed a weak, opaque, chalky-yellow luminescence, the effect being confined to a small area near to the surface. Underneath the yellow fluorescence a reddish-orange fluorescence could be seen. Under LWUV both natural and synthetic alexandrites may show this effect.

Under the microscope parallel striations could be seen along the length of the cabochons. When the iris diaphragm was closed down it could be seen that the striations were undulating rather than straight; they have not been reported from natural alexandrite.

A glass imitating alexandrite was described in the Spring 2004 issue of *Gems & Gemology*. The transparent stone examined weighed 4.45 ct and was a slightly bluish-green in sunlight and equivalent light, changing to purplish-pink under incandescent light. The RI was 1.521 and the SG 2.66. Weak anomalous birefringence could be seen and there was a very weak bluish-green response to LWUV. There was no response to the Chelsea filter. The specimen showed a characteristic rare earth absorption spectrum showing that the colour was due to neodymium.

IMITATION CAT'S EYE

Chatoyant glass cabochons making a good imitation of cat's eye chrysoberyl were reported in the Winter 2003 issue of *Gems & Gemology*.

The glass was made in China and was reported to be common in the gem markets. Other colours had come from the same source, including white-green and green-red-blue as well as different versions of the yellow-brown stones. The article gave the RI range as 1.52–1.58, with orange specimens giving the lowest and pink stones the highest values. The SGs ranged from 2.65 to 3.2 for the same coloured stones. The complete colour list was orange-red, blue-green, 'cobalt' blue, light blue, violet-brown, grey, white, dark green and pink.

Examination of the specimens showed that they were made of parallel glass fibres, which could be seen with a scanning electron microscope. Many of these fibres were not completely straight but slightly bent. Fractures and black spots could be seen in some of the pieces, giving a somewhat natural appearance.

The material was determined as a silicate glass with lead and potassium. Raman spectroscopy showed that it was amorphous, giving a spectrum of two peaks, at about 800 and 290 cm^{-1}. Gemmological testing will show the unusual type of chatoyancy and colour; an optical microscope will show the parallel glass fibres.

Details of the ingenious cat's eye imitation Cathay stone will be found in the chapter on glass.

CHAPTER FIFTEEN

Amber

A darkening of colour noted in some amber specimens over recent years, together with increased RI values and orange fluorescence under LWUV and minute gas bubbles immediately below the surface, has been attributed to heat treatment.

Faceted samples of Baltic amber tested by GIA and reported in the Winter 2002 issue of *Gems & Gemology* ranged in size from 3.2 to 1.3 ct, all with the same yellow colour before heating. Infra-red spectroscopy confirmed that the samples were amber.

With one specimen kept back for reference, the other five were heated up for up to forty-eight hours at temperatures up to 200°C in a nitrogen atmosphere. With increasing duration and temperature the colour of the specimens changed from pale yellow to orange to dark brown.

In the darkest colours the increase in RI correlated with the degree of darkness. Some degree of optical anomaly was observed and the untreated sample had an RI of 1.54. The highest RI was measured at 1.62 (this was the darkest of the samples). One sample gave two readings and another more than ten shadow edges on the refractometer between 1.54 and 1.59. Two other samples showed multiple shadow edges, but only when they were rotated. The variations in RI appeared to correlate with uneven coloration. When these samples were examined on a polariscope with the polarizer only rotating, effects comparable with isotropic materials were seen.

Two samples were cut in half to ascertain the depth of colour. In both cases a brown layer less than 1 mm thick could be seen near the surface. The original pale yellow colour could be seen beneath this layer. The cores of these samples showed an RI of 1.54.

Beetles and other animal inclusions in plastic are too perfect when compared with the dismembered examples commonly seen in amber

Surface colour treatment of amber

The contributor of this descriptive note to *Gems & Gemology* remarked that many of the ambers recently on sale as 'sun-spangled amber', as well as amber with an orange colour, had refractive indexes ranging up to 1.555; this appeared to be consistent with heat treatment.

CHAPTER SIXTEEN

Spinel

S pinel is the name of a group of minerals, all of which belong to the cubic crystal system, but here we are interested primarily in the magnesium spinel with the formula magnesium aluminium oxide, $MgAl_2O_4$. In nature spinel may be found in a number of colours, red being the most valued. The natural stone has an SG of 3.60 and an RI of 1.718. The hardness just exceeds 8.

SYNTHETIC SPINEL

Synthetic spinel differs from artificial materials in that it imitates other gem species rather than itself. Cheaply and easily grown by the Verneuil method of flame fusion, spinel may be grown as boules of almost any colour, although red ones are rare, owing to growth diffi-culties. Probably the variety that gives most trouble is the colourless material so often used as an imitation of diamond, especially small and apparently insignificant ones. Fortunately they give a characteristic response of sky-blue fluorescence under SWUV radiation.

Flame Fusion

The early growth of spinel probably began with Ebelmen, who is best known for his work on the synthesis of emerald. While most synthetic gemstones have the same chemical composition as their natural counterparts, Verneuil growth of spinel needs extra alumina to be successful; the usual composition of the flame-fusion spinel is $Mg2\frac{1}{2}Al_2O_4$, although a successful and convincing imitation of moonstone is sometimes grown with five times the alumina content.

The extra alumina affects the RI and SG, at 1.728 and 3.64 respectively. Like its natural counterpart, synthetic spinel is both hard and durable with no cleavage, although the rare red Verneuil material tends to fragment.

Boules or boule sections of both synthetic corundum and synthetic spinel are sometimes sold at gem and mineral shows. The two can be distinguished from one another by their cross-section, boules of spinel being roughly square and corundum ones rounded. Spinel boules have rougher surfaces than those of corundum, and this is useful when sections only are on offer.

Spinel is a useful material for any kind of ornament. The range of available colours is wide though colourless boules are probably more commonly grown than any other kind. Red is achieved by the addition of chromium and many other colours by iron. Perhaps the commonest varieties of coloured synthetic spinels are green (imitation of peridot) and light blue (imitation of blue zircon).

The early growth of spinel is interesting; it is well known that attempts to grow blue sapphire by adding cobalt to the feed powder used for the Verneuil growth of corundum resulted in the production of spinel instead. This came about because magnesium oxide needed to be added to facilitate the incorporation of the cobalt in such a way that the crystals would not be unevenly coloured. The grower was L. Paris of the Institut Pasteur, Paris, and his report was published in 1908.

This seems to have been an accident from which no commercial developments immediately followed since synthetic spinel did not appear on the market until several years later. Whether spinel is produced in smaller quantities than corundum is uncertain but since natural ruby and blue sapphire are much more expensive than the species imitated by synthetic spinel I would imagine that it is.

Faceted synthetic Verneuil spinel can be distinguished from the gem species that it imitates by its RI and SG. Between crossed polars it provides one of the classic effects of anomalous birefringence: the striped light and dark 'tabby extinction'. Unlike Verneuil-grown corundum there are no bold-edged, randomly occurring gas bubbles and though synthetic spinel occasionally shows curved colour banding there are no growth lines. Bubbles with an unusual shape do sometimes occur, some having been likened to furled umbrellas. Some red boule sections and stones faceted from them show a characteristic Venetian-blind effect but, as I have said, red Verneuil spinel is very rare. When one does turn up, gemmologists may be interested in one feature: in the red end of the absorption spectrum a group of emission lines (bright red) can be seen. From natural spinel these have traditionally been known as 'organ pipes', and in most Verneuil red spinels they are not normally seen, being replaced by a single emission line. However, in some spinel analogues (zinc aluminates doped by chromium to give a red colour) the group of five lines can be seen with particular sharpness. These specimens were probably flux grown rather than Verneuil grown.

I have mentioned that some Verneuil-grown corundum had been coloured blue by doping with cobalt in an early attempt to produce

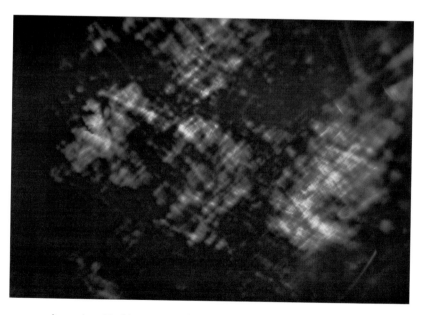

Anomalous birefringence patches characterize a blue synthetic spinel

Verneuil-grown red spinel is rare but most known examples show characteristic curved growth layers

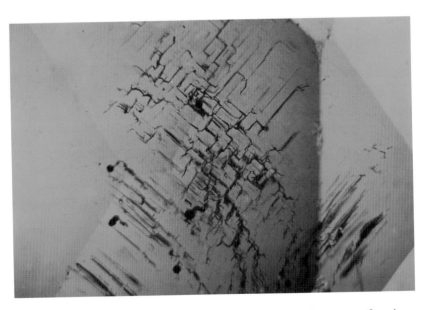

Verneuil spinel (tourmaline green). Spherical gas bubbles and a system of cracks follow the isometric directions of the cubic crystal system

243

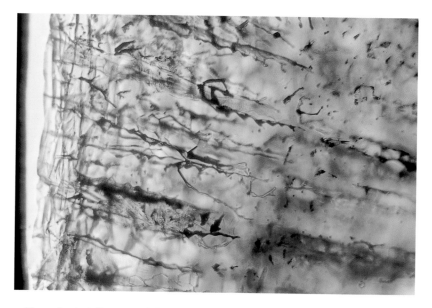

Verneuil spinel (lime green). A dense concentration of fissures and gas-filled 'hoses' (highly irregular formations)

blue sapphire. Magnesium oxide had to be added for the colour to be satisfactory but the final product was not corundum but spinel. However, cobalt has been added to flame-fusion-grown spinel to produce an attractive dark blue which may be offered as corundum. The cobalt allows the crystal to transmit some red as well as the blue and the effect can be seen particularly well when the specimen is strongly lit and viewed through a Chelsea filter (developed for the testing of synthetic emerald). Cobaltian synthetic spinel under these lighting conditions shows a very strong and attractive red through the filter – some specimens will show a red flash when viewed against a white background if the incident light is strong enough. Sometimes daylight will do if the sun is strong. Any red flash from a blue stone should say 'cobalt'.

Blue spinel doped with cobalt will also show a cobalt absorption spectrum rather than a blue sapphire one. The cobalt spectrum, with characteristic absorption bands in the 635–615 nm area, 590–560 nm and between 550 and 535 nm is quite unlike natural blue sapphire's absorption spectrum, whose bands are close together in the blue

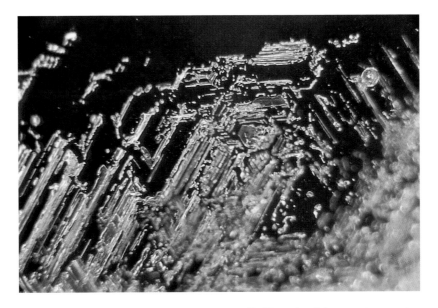

Verneuil spinel (aquamarine blue). Note the gas-filled 'hoses' and the negative crystals

(Verneuil blue sapphire may show only traces of the most persistent band at 450 nm).

As synthetic spinel is so hard and tough it is ideally suited to use in the top section (crown) of a faceted composite whose base (pavilion) may be made of a softer material which on the other hand provides the colour, sometimes another desirable property. Testing may show that the colourless crowns fluoresce a strong sky blue under SWUV; the overall effect of this combined with a base of a different colour would be striking but if this ever occurs it is rare.

Much more dangerous are the complete colourless spinels already described. A piece of jewellery set with a large number of small 'diamonds', when examined under SWUV, may reveal some blue-glowing stones and some which are inert to the radiations, showing no response. This would be the most usual response from such an artefact. On the other hand a similar piece in which all the stones glowed a uniform sky blue would be most unlikely to have diamonds in it. In such a case synthetic spinel is the most likely culprit – but there can be snags if only this test is used. Some synthetic diamonds also fluoresce a sky blue under SWUV, but spec-imens are still less likely to be encountered than examples of the Cape series of (natural) diamonds which also give a similar fluores-cence under the same conditions.

It is too painful to contemplate the existence of a piece of jewellery set variously with natural non-Cape diamonds, Cape diamonds, synthetic diamonds, synthetic spinels and almost anything else. Of course, such a piece would give a varied response; as we have seen, however, a uniform response is highly unlikely to indicate diamonds if many stones are involved.

Although they are rarely seen, we should not forget the chromium-doped specimens which give an alexandrite-like colour change from red to green depending upon whether the light used for viewing is incandescent or fluorescent or daylight (the latter not much use, as specimens are too dark). The manufacturers told me many years ago in New York that the aim of the growers was less to imitate alexandrite than dark green tourmaline. This seems odd – if one wanted to grow cheap 'tourmaline' surely one of the brighter colours would be more saleable. Ironically the spinel 'alexandrites' are much more like the real thing (alexandrite chrysoberyl) than the amethyst-like vanadium-doped synthetic corundum. The RI of the synthetic spinel, at 1.728, is close enough to that of alexandrite (1.74–1.75) for a mistaken reading on the refractometer to be possible.

Flux Growth

Before looking at other versions of spinel we should consider the natural mineral for a moment. Spinel occurs most commonly as small octahedra (equidimensional pyramids joined base to base). These crys-tals are attractive and sought by collectors, to the extent that imitation might be thought worth while.

Until about the 1970s spinel was grown as boules, which of course showed no outward crystal form and which were sold for cutting into faceted stones. I dare say that some boule sections were painstakingly fashioned into octahedra but the only colour in which this would be worthwhile is red; as we have seen, there have always been problems with the Verneuil growth of red spinel.

In the 1970s I was able to examine crystals of spinel which had been grown as octahedra in various colours, including red. The Verneuil method was not indicated; the crystals were in fact flux-grown. It is worth mentioning here that 'spinel' as gemmologists know it means magnesium aluminate: however, minerals with this composition are just one member of the spinel group whose members share similar structures but have different compositions. In some spinel-type minerals the magnesium may be replaced by zinc.

Since the 1970s octahedral crystals of red and blue spinel have been grown in Russia, apparently intended for ornamental use. I have been able to examine specimens of red and blue spinel grown by the flux-melt process by the Russian Academy of Sciences, Novosibirsk. Similar crystals are described in the literature and in particular by GIA in the Summer 1993 issue of *Gems & Gemology*. Both rough and fashioned specimens have been examined.

The colours were strong and well suited to gem use, the blue stones showing red flashes and a saturated medium to dark blue. Some of the blue stones had a higher content of iron than others in the sample tested. Under fluorescent lighting some specimens showed a grey component. The red stones show a medium dark red with an inclination to purple. The RI of 1.717 shown by the iron-rich specimen was in the range for natural spinel and the other specimens also fell in that range. The specific gravity was also consistent with natural spinel, stones sinking in a liquid with SG 3.32. Under LWUV the red stones fluoresced a strong purplish-red to slightly orange-red and the same response, though weaker, was visible under SWUV. Some stones showed slight chalkiness on the edges under SWUV but this is not unusual; a yellowish-orange colour could also be seen in some directions. None of the stones tested showed any phosphorescence.

In the red stones the spectroscope showed an emission line in the red between 685 and 680 nm with a broad absorption band between 560 and 510 nm. The blue stones absorbed from about 450 nm to the end of the visible spectrum. None of the red specimens showed the 'organ pipe' group of emission lines in the red, the single line mentioned above being the only emission feature.

The blue stones gave the characteristic absorption bands of cobalt, absorption being from 635 to 515 nm, from 590 to 560 nm and between 550 to 535 nm. Blue stones showed red to orange through a Chelsea

filter and both red and blue stones showed a transmission luminescence of orange-red colour when viewed in a strong light. Russian
flux-grown spinels showed particles of undigested flux arranged in
net-like formations and isolated pieces with jagged edges. Gas bubbles
could be seen in some of the flux areas. Metallic fragments could well
have been crucible material.

One interesting feature of the Russian flux-grown spinels is the way
in which the larger inclusions of flux form pyramidal shapes which are
arranged along the edges of the octahedron. As natural spinel often
contains chains of very small octahedra there is perhaps a chance of
confusion here.

Colourless flux-grown spinel crystals have been produced from
time to time. Crystals may show triangular markings on the octahedron faces but the two triangles (face edge and surface markings) are
not congruent as they are in natural spinel (a similar feature can be
seen in some flux-grown corundum). These markings (trigons) have
not been reported from the red Russian spinels.

As always the gemmologist has to remember that the absence of
natural solid inclusions is probably the best guide to a tricky synthetic
and this is true of flux-grown spinel. None the less other traditional
gemmological tests are still useful; the blue Russian spinels showed a
stronger cobalt absorption spectrum than the still rare natural cobaltian
Sri Lankan spinel. Under LWUV the blue Russian spinels gave a weak
to moderate chalky reddish-purple fluorescence with the same colour
more strongly seen under SWUV. Verneuil blue spinels may fluoresce
a patchy blue to bluish-white. No response to SWUV is shown by
natural blue spinel.

An interesting use of Verneuil blue spinel is to imitate lapis lazuli.
Here the pyrite commonly seen in lapis is imitated by flecks of gold in
the synthetic spinel. Some pink spinels have been grown, probably by
the flux method.

CHAPTER SEVENTEEN

Glass

The commonest imitation of gemstones is glass. It is cheap, any colour can be made and specimens with a high dispersion can look from a distance quite like diamond. Small stones are the most dangerous.

PROPERTIES

The chemical and physical nature of glass is quite difficult to define. Most glasses from which imitations are fashioned are silica glass (old windows are crown glass, a lime-soda silica glass). The addition of lead oxide gives added powers of dispersion but softens the glass, as well as increasing the likelihood of tarnish and the development of a yellowish colour: lead also raises the SG. Glass containing lead oxide is called flint glass.

Dispersion can also be increased by the addition of thallium oxide and opacifiers can be used to make glass translucent to opaque. Colouring agents (metal oxides) are used for specific imitations.

Cheap imitations of other gemstones are moulded, but the better ones are faceted on a lap in the same way as natural gemstones.

In general glass is relatively soft, with a hardness between 5.5 and 6. It is easily scratched and brittle but shows no cleavage. Facet edges are rounded when the stone has been moulded and conchoidal fractures are especially characteristic. The prominent and well-rounded bubbles can easily be seen with a hand lens and swirl marks are very common and easy to see when the specimen is examined under dark-field illumination (from the side).

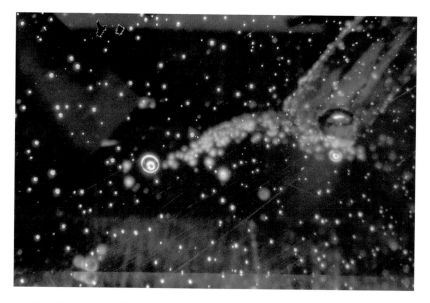

Paste (green glass). Discrete large and small air bubbles are accompanied by a myriad of pinpoint air bubbles

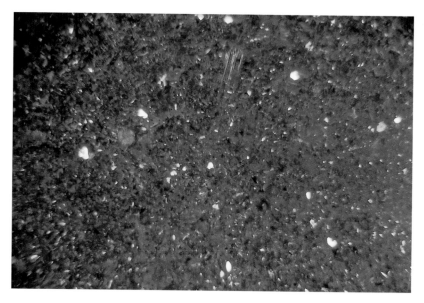

'Goldstone'. A dense accumulation of copper crystals

Glass is a poor conductor of heat and feels warm compared with most natural (crystalline) gemstones.

In general the RI of the glass most commonly used as gemstone imitations will be between 1.50 and 1.60. Most gemstones, and especially those imitated by glass, will be birefringent: glass on its own never shows doubling of inclusions or of back facet edges. SG values cover a wide range, from 2.3 to 5.0.

Glass may also be recognized by its behaviour between crossed polars; it shows apparently irregular birefringence (due to strain), the field being neither completely dark (as an isotropic material ought to be), nor alternately dark and light four times during a complete rotation.

USES IN GEMSTONE IMITATION

Cathay stone is an excellent imitation of cat's eye. Bundles of glass rods are heated and drawn in such a way that light reflected from them when they have been incorporated in a host glass of different refractive index gives an excellent eye. Colouring agents are added to give the honey-yellow chrysoberyl colour. This material is very effective when set and the high constants are not likely to suggest glass in the first instance. The manufacturers, Cathay Corporation of Stamford, Connecticut, USA, give the RI as 1.8, the SG as 4.58 and the hardness as 6.

Glass has been used to imitate star stones, sometimes by having six rays engraved on the back or by using a foil on which rays have been scratched. A white opaque glass has been made to imitate star stones; in one example at least the molten glass was pressed to make a cabochon with six ridges, giving an imitation of the rays of a star. On completion of this process the stone was covered with a thin layer of deep blue glaze which hid the ridges, making it look as if the star was just below the surface.

Opal can be imitated by translucent lime glass with added fluorides or phosphates. Calcium compounds precipitate in the glass.

Victoria stone (kinga-stone or meta-jade) is an opaque partly crystallized glass design to imitate jade, although there is not much resemblance. Fibrous inclusions in parallel bundles may give the effect of chatoyancy.

Iimori stone is also used as a jade imitation but is clear to translucent. Goldstone is made by including cuprous oxide with the glass ingredients and reducing oxygen during annealing. This allows small crystals of copper to precipitate and gives a sunstone (feldspar) effect.

Blue glass has been used as an imitation of lapis lazuli although the glassy effect is too obvious. A green glass, said to have been made from fused material from the eruption of Mount St Helens in Washington, USA, contained 10 per cent at most of this ash.

Glass imitating blue zircon will not show the zircon's strong bire-fringence. When red spinel is imitated the spinel RI of 1.718 will not usually be reached. It will not show the three-band iron absorption spectrum of peridot nor the 'horsetail' inclusions of demantoid garnet.

Devitrification in glass can closely resemble natural foreign inclusions. Crystals of some of the compounds which have been used in its manufacture precipitate and further tests are needed.

B.W. Anderson in various editions of *Gem Testing* gives the SGs and RIs of some coloured glasses in a table. Notable is a colourless glass with an RI of 1.47 and an SG of 2.30. This is a borosilicate crown glass. A yellow calcium crown glass has an RI of 2.43 and an SG of 1.498, and a colourless light flint glass with lead gives 1.54 and 2.87 respectively.

Other examples include an emerald-green fused beryl glass with an RI of 1.516 and an SG of 2.49, and a yellow flint glass, with lead, showing an RI of 1.77 and an SG of 4.98. The dense glass used in a refractometer has an RI of 1.962 and an SG of 6.33 – this is an extra dense flint glass.

Fused beryl glasses would show RIs close to those of natural beryl if they were crystalline and thus denser. Emerald-green specimens show an RI of 1.516 and an SG of 2.49; blue specimens show 1.515 and 2.44 respectively.

Some of the lead glasses have RIs in the topaz range (one glass cited in the table had an RI of 1.633 and was yellow, although its SG was 3.627, above the topaz limit of 3.53).

Glasses with a higher dispersion than most natural gemstones give a sharper shadow-edge reading in white light on a refractometer. These are the lead glasses with RI readings over 1.60. If a spinel refractometer is used the reverse effect is seen, with the lead glass specimen giving a shadow edge with a colour fringe and the natural stone a sharp colour-free shadow edge. This is a good way in which to assess (not measure) dispersion since a specimen with a low dispersion will show a notable colour fringe to the shadow edge on a glass refractometer while a specimen of a high dispersion will not. White light is of course necessary for any colour to be seen and the reverse effect, as above, is seen when a spinel refractometer is used.

A fused quartz with an RI near 1.46 and an SG gravity of 2.21 (natural quartz gives 1.54–1.55 and 2.65) is another example of a glass with a similar composition to a natural material having lower constants because of its amorphous nature. Both the fused beryl and the fused quartz can be coloured by various dopants.

Cobalt is added to give blue and chromium to give green. The cobalt-doped glass shows bright red through a Chelsea filter and shows a characteristic absorption spectrum of three strong broad bands in the red, yellow and green at about 655, 580 and 535 nm. The centre band is usually the narrowest of the three (it is the widest in cobalt-doped synthetic flame-fusion spinel).

Red glasses may be coloured by selenium and then show a single broad absorption band in the green, and some pink or red glasses may owe their colour to rare earths and show a characteristic fine-line absorption spectrum. These are not very common but when pink rather than red they may first attract attention by their colour change from pink to a pale slaty blue.

If a specimen looks fairly normal to the eye but will not give a reading on a refractometer when one might expect to obtain one without difficulty, it will most likely be glass. The difficulty in obtaining a reading from the table facet of many, if not most, faceted or moulded glasses usually means that the table facet being tested is not sufficiently flat for optical contact to be made with the instrument. There are ways of getting round this but after a few attempts, light may well dawn and the specimen be taken to a microscope – where it should have started in the first place.

CHAPTER EIGHTEEN

Composite Stones

Under the general heading of composite stones doublets and triplets are the most likely to cause confusion; there are other types of composite but they are much less often encountered. There are degrees of respectability, too: the opal doublet or triplet can be a way of using exceptionally beautiful material which is too thin to stand on its own as a gemstone and is thus perfectly acceptable in the trade (it would be hard to avoid), so long as its nature is disclosed to the customer.

Doublet. Red reflections along the rim of the crown and fractures below the girdle betray this doublet with an almandine crown and glass pavilion

Doublet (imitating emerald). Lateral view showing the beryl crown and green glass pavilion

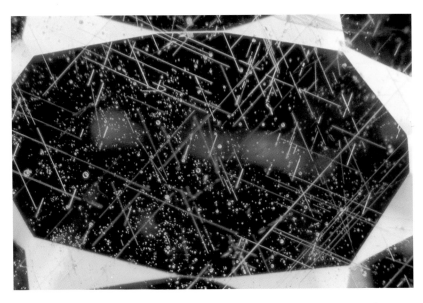

Doublet (imitating emerald). Rutile needles in the almandine crown are in focus. Air bubbles in the cement layer are blurred

While the names would appear to be obvious, none the less they are worth defining. A doublet consists of two sections joined together in such a way that they cannot separate easily under normal conditions of wear. Composites are hardly ever actually made from portions of the stone species that they are intended to imitate. Crown and pavilion are the upper and lower sections of the normal round brilliant or rectangular step-cut stone. The crown is the area upon which testing is likely to centre and so should be the harder part of the composite. A hard, commonly available material should be chosen: almandine garnet most often meets the case and the term 'garnet-topped doublet' is used for a composite consisting of a thin layer of almandine fused to a brilliant or step-cut stone which is most commonly a much softer glass. The body colour of the composite is provided by the glass. Students often represent a garnet-topped doublet as joined at the girdle but the slice of garnet has to be thin enough to avoid making the whole piece red (although there are red garnet-topped doublets consisting mostly of red glass).

A triplet consists of three pieces and is most often found with a colourless or pale crown and pavilion, and a glass of the same but

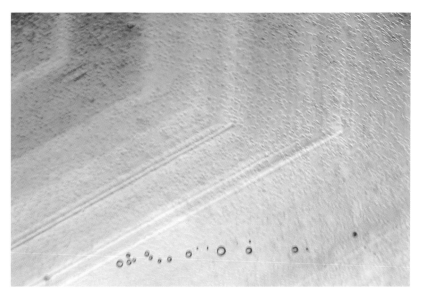

Doublet (imitating blue sapphire). Hexagonal zoning in the natural sapphire pavilion. Gas bubbles in the synthetic sapphire crown

Doublet (imitating whole natural ruby). Straight parallel lines (edges of polysynthetic twin lamellae) in the natural ruby crown. Spherical air bubbles in the cement layers above the synthetic ruby pavilion

'Naftule' doublet. Synthetic colourless corundum crown is attached to the strontium titanate pavilion below the girdle. Note the fine separation line

deeper colour at the join, which is normally at the girdle. Some triplets are coloured only by a dyed cement at the girdle.

It is hard to find out who is making composites. Nobody seems to advertise them and they have been on the scene for well over 100 years. This is a task for a gemmological detective.

GARNET-TOPPED DOUBLET

Textbooks usually say that one of the reasons for the development of the garnet-topped doublet was that jewellers in the past were believed to draw a steel file across the table of a specimen suspected of being soft glass, which would be scratched. Customers would have objected if a lot of glass was damaged in this way, and since easily available, a thin slice of garnet was fused to avoid this. However, this would take valuable time and one can assume that if the practice was really widespread specialist doublet-makers would have been recorded.

I have said many times that in testing any stone, one should 'think composite' sooner rather than later. Many composites show 'odd' colours to the experienced eye and if they are garnet-topped doublets (whose garnet top has too high a refractive index to be tested on a standard refractometer) the specimen will not give the reading expected for the stone it is pretending to be – ruby or emerald as well as peridot, zircon, topaz or tourmaline.

The reason for the odd colour is the slight amount of red passed by the garnet slice. We have to remember that as red garnets go, the iron-rich almandine is very dark and even a thin slice is red enough to alter the rest of the colour, provided by the glass base, in such a way as to arouse suspicion. As soon as it is aroused, the stone should be held under a strong source of white light (fibre-optic sources are ideal), and will show red flashes as it is moved.

If red flashes are suspected they can usually be proved not to have been imaginary by placing the stone table down on a sheet of white paper. A red ring will then be seen round the table (or rectangle with a step-cut stone). Students should practise on known garnet-topped doublets to get some idea of what to look for. This is important when the garnet-topped doublet is fused to a red glass body to imitate ruby.

The gemmologist will then have to make a distinction between two kinds of red, neither of which will be very like a true ruby.

What happens when a red flash or red outline of the table is not detected? The stone should be tested by the normal methods. If it is a transparent green (in the hope that it will be accepted as emerald), a spectroscope would be a reasonable choice for the next test. The green glass will give no sign of the characteristic emerald absorption spectrum but some of the iron bands associated with almandine might well show up if the specimen is examined in a suitable orientation, with the spectroscope picking up the light transmitted from the garnet slice. The persistent strong band at 505 nm will certainly make its presence felt and this should rule out emerald from the outset.

Although most emeralds do not fluoresce very easily under LWUV (despite what some textbooks say) the glass and garnet certainly will not. More significantly, a Chelsea filter, through which all but iron-rich emeralds will show a fairly bright red, will do nothing for a garnet-topped doublet.

On a refractometer the RI of 1.81 will not be seen with today's contact liquid with n=1.79, but nor will the RI expected for the emerald or other imitated species.

If the first instrument used is a microscope then the characteristic inclusions of glass will be seen in most of the stone. These will be well rounded and randomly distributed gas bubbles and a swirliness which is hard to describe but when observed a number of times comes to be familiar. Examination of the facet edges will show numerous conchoidal fractures, unless the specimen is fresh from the manufacturer.

With a red doublet a spectroscope will not show the chromium elements that would be easy to see in ruby and red spinel. The 505 nm absorption band will give a clue when the glass will show only red being transmitted – in a deep red specimen the red will stand out against a nondescript dark background.

Assuming that a red stone will attract more investigation than stones of most other colours, a microscope could well be used. If the garnet layer is examined, one may be able to see the parallel groups of acicular crystals of the titanium oxide rutile, which may form a reticulated structure.

It is possible that the glass portion of the doublet may show a white or pale green glow under SWUV and if the RI is taken (hardly worth the trouble bearing in mind that a small facet will have to be tested) the reading will be somewhere between 1.62 and 1.69, although lower readings have sometimes been obtained.

Even if a specimen has been proved to be a garnet-topped doublet by other methods of diagnosis, it is still worth taking time to examine them in case some new method of manufacture has been used. For example, although garnet is chosen for its easy availability and because it is one of the few substances that will fuse to glass, there is always a possibility that some other substance will be tested or that garnet may still be used but in a different way, perhaps extending evenly further down the upper facets of the stone instead of the often uneven thin slice at the very top.

Immersion, so useful in a number of ways, is also of value in the identification of a doublet, but one has to remember that the two portions might come apart under the influence of the liquid selected – although this is less likely with a fused garnet-topped doublet. And while immersion will certainly show up the internal characteristics of glass it will also show the differences in refractive index between the glass and the garnet layer – they will appear quite different in relief.

In *Webster's Gems*, 5th edition, 1994, John Rouse cites an Austrian book published in 1926, Albrecht Schrauf's *Handbuch der Edelsteinkunde* as giving useful information on the fusing of the two portions of a doublet. Another book, reported by Rouse to be by G. Burdet and entitled *L'industrie lapidaire* (which may be his *Etude historique sur la pénétration et le développement de l'industrie lapidaire sur le plateau de Septmoncel et dans la région de St-Claude*, published at Morez-du-Jura in 1925) claims that the garnet-topped doublet was invented in 1845 by Cartier, lapidary to the Comte de Mijoux.

Rouse also quotes from an account by Frank E. Goldie of his life as an apprentice in the Jura area, presumably of Switzerland, during which he assisted a family in the manufacture of garnet-topped doublets:

> After the day's work … a small pile of doublet moulds [was] put on the kitchen table, together with a packet or two of thin slices of garnet and also a mound of glass squares. Everyone sat down to their respective job … Some put in the garnet, others would place a piece of glass on top. The moulds were made of baked

clay, measuring about 16″ by 10″ with a number of indentations depending on the size of doublet required … Friday was an important day, the kiln was stacked with prepared moulds, the fire lit and heated to the correct temperature. Next morning, the moulds having cooled were removed. The rough doublets were then sorted into their different colours and sizes ready to be given to the local craftsmen who cut them in their homes.

DIAMOND DOUBLETS

Here is one of the rarest of all specimens. The diamond doublet is sometimes placed in a group once known as 'genuine' doublets, an unnecessary name which has now lapsed into disuse. Simply, a diamond doublet is composed of two diamonds, both faceted and joined together at the girdle by some kind of cement. The clue to identification is inside: viewed through the table facet another table seems to be visible some way down inside the stone. A pencil-point placed on the table can be seen repeated on the second table – this is a very easy test, and the result is fascinating to observe.

It is perfectly possible to assemble doublets of this kind in any material but it is difficult to imagine that it would be profitable, except perhaps with corundum, of which there are reported examples.

OPAL DOUBLETS AND TRIPLETS

These are beyond doubt the most important composites encountered by the gemmologist, since opal is a valuable material and its use in composites is by no means always intended to deceive.

Genuine opal doublets are always made from opal, and no other material is involved. Quite simply, opal occurs so often as very thin seams that it would not be able to withstand normal wear unless backed by some harder material. The most convenient, and to some the most genuine answer to this problem (and it *is* a problem since some of the thinnest opal shows the best play of colour) is to use the rock (matrix) on which the opal seam occurs. This is exactly like making a stone from the butter on bread and leaving the bread for support. In some cases opal is left as it is on the matrix and sold as opal matrix – a very attractive material – but this is nothing to do with composites.

The best and most immediate clue to an opal doublet is the generally flat appearance. Looking at the side of the stone should reveal the change in the play of colour which stops suddenly at the junction between the opal and the backing. Naturally the setting will often make examination difficult.

It may be possible to pass a strong beam of light through the stone, in which case air bubbles in the junction layer may become visible. The general thinness of the stone will help here. This test will not be possible with a closed setting, and in this case unless the opal is a particularly good specimen, testing would not be profitable in terms of the time spent. In the case of really fine opal in a closed setting there is nothing for it but to remove the stone for examination.

An opal doublet can be made into a triplet by capping the opal layer on its backing. The best opal triplets are capped by rock crystal but these seem to be quite rare. Glass caps are known but plastics seem to be much the most commonly used. Both glass and plastics are quite easily scratched. A triplet will appear less flat and more domed than a doublet; the domed cap, whatever material it is made of, acts as a convex lens and so magnifies the play of colour as well as protecting the relatively fragile opal layer.

It should be said that doublets and triplets can be made using synthetic opal (see the chapter on opal). Seeing an obvious doublet or triplet, one may well assume that the opal must be natural but the colour patches should be examined to see whether there are the hexagonal outlines characteristic of synthetic opal.

Rather more elaborate tests could be made on a suspected triplet. The quartz (rock crystal) cap might show a uniaxial ('bulls eye') interference figure between crossed polars on a polariscope and it might be possible to take the RI of the quartz. If the stone is free from its setting the base will of course be available for different kinds of test. It should be examined from the side as a matter of routine.

Plastic caps might be tested by a thermal reaction tester in search of the characteristic plastic smell, but resorting to this kind of test is surely clutching at straws.

It is worth noting that a spectacular black opal, by far the most highly prized variety, should be examined with extra care since the type of low-domed triplet with the base made from black onyx and the cap from rock crystal can sway the imagination.

An opal doublet with the play of colour against a milky-white background and cemented to a base of dark blue sodalite was reported in the Summer 1996 issue of *Gems & Gemology*. No doubt the dark blue base would have given a pleasing and durable background to the play of colour, perhaps strengthening the blues.

Some opals (the examples I have seen come from Idaho and have a very pleasing play of colour) show the star effect but are clearly composites, with no attempt to hide their structure. The original opal is chosen for a band pattern which is made more like the ray of a star when capped by a cabochon-shaped cap, usually of some form of plastic, which acts as a condensing lens. The flat plastic base leaves one in no doubt as to its nature. These are attractive stones and seem to be surprisingly rare.

A black opal triplet described in the Fall 1989 issue of *Gems & Gemology* consisted of a natural flat-topped opal with an uneven lower surface, giving a wavy effect. The backing was fashioned from ironstone and joined to the lower surface of the opal using a cement tinted to resemble the appearance and colour of the ironstone. The uneven contact surface of the opal top was filled by the cement to give the appearance of a single stone. When lightly touched with a thermal reaction tester the joining material melted.

The name opalite has been given to an opal triplet in which the play of colour is provided by a mosaic which is covered by a clear top and has a wax-like base. The mosaic is made up of flat pieces of natural opal and in some cases the adhesive binding them phosphoresce. The name 'opalite' has been used elsewhere and will be again in time to come, I am sure.

COMPOSITE EMERALD IMITATIONS

After opal, emerald is the major gem species to be imitated by composites. The name *soudé* has been used rather indiscriminately but means 'sandwich' – a stone in which the colour is carried by a central zone usually corresponding with the girdle. Almost any kind of composite could be made to imitate emerald: two pieces of green glass cemented together, two pieces of colourless quartz (rock crystal) with the colour coming from a green glass cemented between the two or a synthetic emerald overgrowth on a cut seed plate of natural colourless beryl.

It is impossible to overestimate the kinds of composite that have been made at some time or another. In almost all cases, I should imagine, no money has been made but scientific and engineering interests and instincts have probably been satisfied. Those engaged in the testing of gemstones should therefore never rule out anything.

The variety of emerald known as *trapiche* from its supposed resemblance to a type of cog wheel used in the harvesting of sugar cane is very popular with collectors. It has been found so far only in Colombia and was first reported in serious literature in the early 1970s. The intriguing crystals have an emerald core surrounded by segments of emerald arranged on the faces of the hexagonal crystal. *Gemmologie*, 51 (2002), described and illustrated an imitation in which both the central core and surrounding segments were emerald, the only difference from the natural material being that the segments had been glued on. As so often, we wonder why, and how many examples are around. This is a different kind of composite.

OTHER COMPOSITES

The material known unofficially as goodletite, found in the South Island of New Zealand, shows an attractive range of colours, including ruby red and emerald green. It has also been used as the basis of a composite rather in the same way that ammolite has; it is too thin and fragile to stand alone and needs to be capped, usually with plastic.

Some of the ammolite triplets may show traces of the cementing substance and the cracking between the segments (which gives the general impression of tortoiseshell on a smaller scale) should be a good aid to identification.

Tourmaline has also been used as a doublet, either with glass or with some form of plastic – examples of these and of some of the other types I have mentioned can be found in *Gemmologie* 51 (2002).

The Winter 2003 issue of *Gems & Gemmology* summarized as an abstract from *Australian Gemmologist*, 21(6) 2002, a list of the more frequently found composites turning up currently in Germany:

- doublets and triplets imitating the more important gemstone varieties, diamond, ruby, sapphire and emerald
- doublets and triplets imitating 'trendy' colours or attractive stones recently discovered; an imitation of blue Paraiba tourmaline has a crown of colourless tourmaline and a glass base carrying the colour
- composites in which fragile material is protected by a harder cap, as for ammolite, using glass or synthetic spinel
- composites to protect very thin ornamental layers as in opal
- composites in which the ornamental material is attractive only in thin layers (a thin crystal section of zoned tourmaline backed by colourless glass)

One interesting example described was an imitation of cat's eye aquamarine, consisting of a crown of beryl and a pavilion of chalcedony. Another was a moonstone imitation made by cementing a rock crystal crown to an orthoclase feldspar base, the latter providing the moonstone effect.

CHAPTER NINETEEN

Less Common Synthetics

LANGASITE

Langasite is an orange-to-yellow synthetic gem-quality material, grown in the United States, Russia and elsewhere. Probably grown first for its piezoelectric properties, with the composition $La_3Ga_5SiO_{14}$, it has not been found in nature. It is a member of the trigonal crystal system. Faceted gemstones are described and illustrated in the Fall 2001 issue of *Gems & Gemology*, the photograph showing two faceted stones of 56.80 and 4.59 ct.

Specimens tested by GIA showed uniform coloration with no banding and a moderate pleochroism of light yellow-orange. The SG was measured hydrothermally at 4.65; the RI could not be read on a gemmological refractometer but was measured at 1.909 and 1.921 for the ordinary and extraordinary rays respectively. The birefringence was found to be 0.011 and the dispersion 0.035. The hardness was about 6.6.

No specimen showed any trace of fracture or of twinning. Triangular-shaped solid inclusions in linear arrays could be seen in one sample. Rough samples with crystal faces showed growth striations in directions oblique to the direction of the optic axis. The samples in the test were found to be inert to both types of UV radiation.

It was not possible to see any absorption in the visible with a hand spectroscope and there was no transmission luminescence (this may sometimes be seen when a specimen is strongly illuminated by white light). The samples showed no fluorescence under either type of UV.

Five orange-to-yellow specimens showed increasing absorption from about 580 to 350 nm with no absorption from about 850 to 580 nm.

The infra-red absorption spectrum contained a broad band centred at 5,436 cm⁻¹ with weak, sharper bands at 3,405, 2,921 and 2847 cm⁻¹. Raman spectroscopy gave peaks at 5,280, 5,246, 3,472 and 3,223 cm⁻¹, with a broad peak centred at 4,612 cm⁻¹ and weak peaks at 3,574, 2,654, 2,628, 737 and 704 cm⁻¹.

Examination by energy-dispersive X-ray fluorescence (EDXRF) showed that lanthanum, gallium and silicon were present but no trace elements could be detected. Heating the sample to 250°C did not affect the colour, whose origin has not yet been determined.

In general langasite should be distinguished from diamond by its birefringence and from corundum by its higher RI. It could perhaps be confused with zircon but the latter shows an absorption spectrum which is easy to recognize.

In the same issue of *Gems & Gemology* a laboratory-grown pink material with the composition $SrLaGa_3O_7$, and thus related to langasite, was described. The sample tested by GIA was a light pink in daylight, changing to light yellow when exposed to strong incandescent light. The change in colour occurred slowly, unlike other colour-change species such as alexandrite, in which the change is immediate. The RI was found to be near 1.82; between crossed polars a moderate to strong birefringence could be seen, although this could not easily be read on a refractometer as the shadow edge of 1.82 was too close to that of the contact liquid.

The uniaxial sample appeared to show no dichroism but GIA considered that the light colour of the specimen tested might have been responsible for the apparent absence of this effect. The SG was high, at about 5.25. The stone fluoresced an intense green under LWUV and a weaker green under SWUV.

The absorption spectrum as shown by a hand spectroscope consisted of a doublet in the green near 525 nm and a line in the blue at about 485 nm. Groups of inclusions included whitish breadcrumbs and short cylindrical tubes with interruptions in some places. Some of the tubes appeared to emanate from some of the larger breadcrumbs. Some inclusions appeared to resemble negative crystals and there were a few small, irregular yellowish masses.

An energy-dispersive X-ray spectrometer showed that the composition of this material was very close to that of langasite. The cause of the pink colour was investigated by obtaining a visible-UV absorption spectrum which consisted of a number of sharp lines and doublets which equated with those shown by erbium, found as a dopant in some examples of cubic zirconia.

FORTALL

In June 2003 *Gemmology Queensland* reported a new emerald-green synthetic material marketed through Rough Synthetic Stones of Bangkok (the original report appeared in *Jewellery News Asia*, May 2003). The stones are sold under the name Fortall.

The material was said to be grown by crystal pulling and had a hardness of 7, an RI of 1.72 (the material was believed to be singly refractive), and an SG of 2.70.

As well as the emerald-green variety, black, blue, brown and red colours were said to be available as either rough or cut specimens. It was first used industrially.

TANZANITE

Tanzanite, possibly diffusion-treated (there is no synthetic tanzanite) was reported in the Summer 2003 issue of *Gems & Gemology*. Two stones were shown to GIA's New York laboratory. The specimens were a deep purplish blue and weighed 4.19 and 2.51 ct. One was mixed cut and the other cut into a triangular shape.

No signs of fractures or inclusions could be seen with a gemmological microscope and the colour did not appear to be confined to the surface area, as is common with the majority of diffusion-treated stones, when examined under immersion.

One of the stones was sliced through the centre so that chemical analysis could be more easily carried out, using an electron microprobe. This showed that an area near the culet was slightly lighter than the remainder of the stone and straight colour banding could be seen when the stone was immersed.

A glass intended to resemble tanzanite is described in the Winter 2003 issue of *Gems & Gemology*. The faceted violet-blue stone weighed 5.42 ct and showed a single RI of 1.700, the same as the higher value of

RI for tanzanite. No birefringence could be detected. Most glass imitations of tanzanite previously examined had showed an RI near 1.66. Weak anomalous double refraction was observed in this specimen. The hand spectroscope showed weak bands near 600 and 500 nm.

No particular inclusions could be seen and the SG was measured at 4.11. Under LWUV the stone gave a weak blue fluorescence and a weak to moderate chalky yellow and blue under SWUV, FTIR and Raman spectroscopy gave results consistent with artificial glass; EDXRF showed the presence of silicon as the main element with some calcium, zinc, strontium, barium, zirconium, antimony and lanthanum.

RAINBOW CALSILICA

A man-made material which has been named Rainbow Calsilica was reported in international issue no. 10 of *SSEF Facette*, 2003. A necklace of banded opaque stones was shown at the 2002 Ste Marie Aux Mines Mineral Show in France. Described by the vendors as a microcrystalline calcite bonded with the amorphous clay mineral allophane, the two stones tested by SSEF (the Swiss Gemmological Institute) appeared to show multi-coloured banding to the eye, while under magnification several differently coloured layers could be seen, including blue, several shades of green, reddish-brown, white and black. The layers contained differently coloured small grains, interspersed with larger, partly idiomorphic white grains of various sizes. A soft transparent material surrounded the larger grains and also occurred in between the layers. Some of this material contained bubbles.

Raman spectroscopy identified the white grains as calcite. The coloured grains, tested by the same method, did not give results corresponding with those recorded for copper minerals but came closer to the Raman spectra of artificial colours as seen in dyed coral.

Further enquiries showed that the Raman spectra of the dark blue veins corresponded to the spectrum given by twentieth-century blue artists' pigments. The material could be identified as the copper phtalocyanine pigment PB15, a synthetic pigment developed in the 1930s. The light greenish-yellow areas contained the yellow mon-azo-pigment PY1 (Hansa yellow).

The soft, transparent plastic-like material was closely related to paraffin wax. It was thought likely that it was an aliphatic polymer or a paraffin wax derivative mixed with other unknown compounds.

Hematite was the cause of colour in the red bands and the strontium sulphate celestine was identified in black areas and grains. The black areas, examined under 50x magnification, were found to be a dark greyish-green with many different grains, transparent polymer-like layers and dyes.

SSEF noted that the identification of these different materials marked the beginning of a wider use of Raman spectroscopy in areas other than inorganic material testing. Though Raman spectroscopy is not and has never been the sole preserve of mineralogists and gemmologists, its use in an area where organic and inorganic unknowns are combined suggests very interesting possibilities in the future.

TOURMALINE

Tourmaline with an iridescent coating was described in the Fall 2003 issue of *Gems & Gemology*. Near-colourless stones from Nigeria were sent by an American dealer to Azotic Coating Technology Inc of Rochester, Minnesota, USA. The firm was able to produce a range of surface colours on the pavilions of faceted stones and on the domes of cabochons. The firm stated that similar coatings have been applied to topaz, beryl, cubic zirconia, sapphire, quartz and various glasses, as well as tourmaline which in the illustration was shown with pink, yellow and blue colours; the firm stated that greenish-blue, purplish-pink, brownish-orange, greyish-purple, greenish-yellow and orange were available.

The stones gave RIs (observations taken on areas where there was no coating) in the range of 1.640–1.642 with a birefringence of 0.014–0.016. The SG was 2.06–3.09. The lustre of the iridescent coating was described as sub-metallic; no response to either kind of UV was shown by any of the specimens, nor was any absorption spectrum observed in the visible.

One stone gave a vague RI near 1.70 on the coated area and a purplish-pink stone gave a shadow edge around 1.63 with another one near 1.77. This might lead to the stone being confused with topaz, which is more often coated.

CHAPTER TWENTY

Plastics, Resins and Metals

PLASTICS AND RESINS

Plastics have become collectable in their own right and while they do not provide the best imitation of faceted gemstones they can test the gemmologist's skill when offered as amber or even pearl, although the lightness of most plastics means that a necklace will not hang evenly. Maggie Campbell Pedersen in *Gem and Ornamental Materials of Organic Origin* clears up some questions of nomenclature when she suggests that the term polymer could and perhaps should be used in place of plastic, since a polymer can be softened and moulded. The term thermosets has been used for those polymers which, once cooled after formation, remain unaffected by any further heating. Those polymers which can be reheated and reshaped are called thermoplastics.

Amber is a fossil resin, and the passage of time may gradually allow a contemporary resin like copal to take on the hardness of amber, but this transformation cannot be accomplished by heating. Some manmade resins also lose their plasticity over time so that moulding becomes impossible.

Plastics as imitations of amber, coral, shell, tortoiseshell and ivory include the highly inflammable cellulose nitrate, which can be dyed to give most required colours. Cellulose acetate, a later development, is not inflammable and does not degrade with time to give the camphor smell that is characteristic of cellulose nitrate. Cellulose acetate may give a vinegary smell in some circumstances.

The milk-based plastic casein can be dyed to give a range of attractive colours and is generally stable. Campbell Pedersen makes the useful point that casein as an imitation of ivory could lead to confusion as both materials fluoresce a similar light sky-blue under LWUV. The trade names Erinoid and Galalith have been used for casein.

Perhaps the most often used plastic in the ornamental materials context is Bakelite, which was first patented in 1907. The phenol formaldehyde may darken with time, although dyeing is possible. It makes a good amber or jet imitation. Bakelite has been dyed red to imitate Burmese amber and the name cherry amber has been used in this connection. Bakelite does not degrade with time and Campbell Pedersen quotes reports that dyed pressed amber has been offered as Bakelite. This is not as unlikely as it sounds: early plastics have their own following and can realize significant prices.

While the materials described above had been around for many years, more modern plastics have been used to imitate natural ornamental materials. They have also been used, usually by one-person designers and manufacturers, for their own sake, as they can be very beautiful.

Perspex is an acrylic resin which can be made glass-like and clear for use as the cores of imitation pearls and for cheap beads. The RI is near 1.50 (the usual value for plastics) and the SG 1.18, low for plastics.

Polystyrene resins have also been moulded to resemble faceted stones. They have an RI of 1.59 and an SG of 1.05. They are sectile and will dissolve in some organic liquids (toluene and bromoform, once used in testing, are now proscribed for health and safety reasons).

Fire opal can be imitated by plastics, although opal has a higher SG (around 2.00) than most plastics. A brine solution will distinguish amber (SG 1.08) from most plastic imitations (usual SG around 1.05–1.55).

Most plastics can be placed in the 1.5–3.0 area of Mohs' scale of hardness, and a thermal reaction tester will produce an acrid smell. A general RI range for plastics would be 1.5–1.6. All the values will be slightly increased if the plastics contain fillers.

Most plastics will show some sign of degradation over time, smell being an early warning. Some specimens will crack or craze and others discolour. Campbell Pedersen recommends keeping specimens loosely wrapped in acid-free paper: airtight storage is not recommended, nor is storage close to heat sources or chemical products. They should not be exposed to bright sunlight and if signs of degradation appear, they should be kept apart from other plastics.

METALS

The use of metals to imitate natural ornamental materials presents few problems except in the case of the imitation of hematite (which is the alpha form of iron oxide with an SG of about 5.1) by a mixture of stainless steel with chromium, lead and nickel sulphides. Natural hematite, when drawn across a streak plate (made of unglazed porcelain) leaves a streak of red powder; the imitation, which has been called hemetine, leaves a black streak. Most natural hematite does not respond to the pull of a pocket magnet; hemetine is attracted. Hemetine has an SG of nearly 7 and the hardness is near to that of hematite.

None of the imitations of hematite give a red streak and some have a much lower specific gravity. One imitation gives an SG of 2.33. The colour is bluish-grey and the streak dark brown. Natural hematite and its imitations are quite widely used in cheap jewellery.

A powdered lead sulphide which may contain added silver, has an SG between 6.5 and 7 and a hardness between 2 and 3. The material fuses easily and is notably brittle.

A hematite imitation made from titanium dioxide has a yellow-brown streak with a steel-grey body colour. The SG is nearly 4 and the hardness 5.5. Another possible imitation of hematite could be the black form of the synthetic garnet yttrium iron oxide. This would also not give a red streak.

CHAPTER TWENTY-ONE

Collectable Synthetics

This chapter is concerned with some rarer but still highly interesting materials which may turn up, not perhaps in jewellery but in mineral shows where collectors gather. Many of these rarer products were named for the first time with references in my *Synthetic Gem Materials*.

Lithium niobate can be colourless or doped to give a range of colours, one of the more startling ones being a bright violet. Its hardness is over 5, RI 2.21–2.30, birefringence 0.090 and SG 4.64–4.66. The colourless version was offered (not widely, in all probability) as an imitation of diamond and since the dispersion is 0.130 (three times that of diamond) the stones are worth examining. The low hardness and high birefringence rule out diamond when specimens are tested. Linobate was one of the trade names used.

Lithium tantalate is colourless, with a hardness of 5.5, RI 2.175–2.22, SG 7.3–7.5 and dispersion 0.087 (twice that of diamond).

Yttrium aluminate is also colourless or doped to give a wide range of colours. It is much less common than the synthetic garnets, with a hardness of 8.5, a single RI of 1.94–1.97, an SG of 5.35, and dispersion 0.033. Some doped specimens show a rare earth absorption spectrum.

Another singly refractive material also containing yttrium was given the trade name *Yttralox*. It has a hardness of 6.5–7, RI 1.92, SG 4.84, dispersion 0.050. When grown this material is colourless but reports suggest that some specimens turn yellow over time, when the composition varies from the ideal. This would explain its scarcity, as alteration would probably not serve its original research purposes.

Hard, highly dispersive materials may well be taken for diamond if there is only a single refractive index, but one or other of the diamond reflectivity meters or testers should make the distinction easy.

While cubic zirconia (CZ) is well established as today's most successful diamond simulant, the analogous material *hafnia* (hafnium oxide) has also been grown in colourless transparent form. It has not come to be used ornamentally, perhaps because it is more expensive to produce on a sufficiently large scale. There is no trade name.

The fact that germanium and silicon have some properties in common has led to the growth of germanates, in particular *bismuth germanate*. There is more than one possible composition. When transparent and faceted, stones are spectacular, with high dispersion. Most are soft, however, with a hardness of about 4.5. The single RI is 2.07 and the SG 7.12. The body colour is a bright golden orange. As usual a reflectivity meter will show that specimens are not diamond and in any case the very high SG will make faceted stones notably heavy. I have seen only large faceted specimens. A *bismuth silicate* has been grown in colourless and in orange to brown forms. No trade names are reported.

Bromellite has been faceted. This is surprising because the dust from this beryllium oxide is highly toxic. Stones are colourless, with a hardness of 8–9, with RI 1.720–1.735, birefringence 0.015 and SG 3.0–3.02. There may be a weak orange fluorescence under LWUV.

Phenakite, the beryllium silicate, has been grown for experimental purposes. Doping with vanadium gives an attractive light blue and crystals are slender, making them collectable if they can be found. Some of the specimens I have seen have small well-shaped crystals growing from some of the larger faces.

We have seen that natural *zincite* is a fine red; it has been grown by the hydrothermal method to produce colourless, orange, yellow and pale green varieties. At the time that my *Synthetic, Imitation and Treated Gemstones* was published, some large, clean transparent faceted zincites appeared in a red-orange and a green form. The stones were claimed to be natural but it was perhaps coincidental that a few stones came on to the market at a time when the synthesis was proceeding. The supply of stones appeared to dry up after a short time so there is no certainty about their true origin.

Scheelite has been grown on a fairly large scale for industrial purposes and some crystals have been cut. They fluoresce a strong sky blue under SWUV as do their natural counterparts. Some crystals have been doped to give different colours but the strong birefringence rules out a possible diamond imitation. Neodymium doping has produced a purple specimen and this shows the usual rare earth spectrum with two groups of fine lines in the yellow and in the green.

Despite its low hardness (4) and easy octahedral cleavage, crystals of *fluorite* have been grown for purposes other than ornament and some have found themselves faceted. The properties are the same as for the natural mineral but doping has produced some rare colours; one is a red in which uranium is reported to be the dopant. One red uranium-doped fluorite crystal showed a fine line absorption spectrum in which the strongest line was at 365 nm (outside the visible region). A brilliant green fluorite gave an exceptionally long phosphorescence after X-ray irradiation – the dopant was reported to be indium. Another red fluorite was found to contain a number of cavities, straight growth planes and crystallites, the combination giving a deceptively natural appearance.

Periclase, magnesium oxide (MgO), in colourless faceted form has been marketed under the name Lavernite. Specimens have a single RI of 1.73 and an SG of 3.5–3.6. Some stones show a whitish glow under UV. Irradiation has turned some specimens blue, dark blue or green. The hardness is 5–6.

Greenockite, the transparent orange cadmium sulphide, is grown by the vapour-phase method to give quite large crystals and faceted stones which have a hardness of 3–4, an RI of 2.50–2.52 and an SG of 4.7–4.9. The faceted orange stones have too high a refractive index to look like orange garnets, still less fire opal, but they are attractive. This is quite a rare material.

Strontium titanate (whose trade name is Fabulite, although I have not seen or heard it for some years, presumably since the arrival of synthetic garnets and then of cubic zirconia) is grown by the flame-fusion method, so faceted stones are cut from the boule. Completely colourless stones can be obtained. The dispersion, at 0.190 (about four times that of diamond) and the clarity make stones very dangerous in small sizes and as the base in composites, when the low hardness of 5–6 is offset by the use of a colourless synthetic corundum or synthetic spinel top. There is quite a good resemblance to diamond if one is not familiar with diamond's almost inevitable mineral inclusions. The single RI of 2.40–2.41 is very close to diamond's 2.42 and the SG, at 5.13, is immediately felt to be greater than diamond's 3.52.

Strontium titanate can be doped to give interesting colours which certainly could be mistaken for those of coloured diamonds with which customers may not be familiar and which in any case vary widely. On the whole the range of possible colours achieved from doping runs from deep red or reddish-brown through orange and yellow to blue, purple and black. The lighter colours are the most likely to be mistaken for those of diamond. Gas bubbles can be seen in specimens, as with all flame-fusion stones.

With *rutile*, stones are so strongly birefringent (0.287) that diamond could only be considered by those quite unfamiliar with it. Stones are never completely white and while this may suggest Cape diamonds the colour is not quite the same, and in any case no absorption can be seen at 415.5 nm. In addition the dispersion is so high, at 0.28–0.30, about six times that of diamond, that rather than showing diamond's flashes of colour as the stone is moved under a spotlight, rutile's dispersion is more like the play of colour in opal. The hardness is 6.7 and the RI 2.61–2.90, unmeasurable on a gemmological refractometer; the SG is 4.25. An absorption band at 425 nm acts as a cut-off to the spectrum in the violet.

Light blue colours are achieved by oxidation after growth. The addition of cobalt or nickel, without oxidation, produces red, amber to yellow. Red can be achieved by the addition of vanadium and chromium: beryllium gives a bluish-white. A less yellow cast can be obtained by the addition of aluminium to the feed powder. Star stones are made by adding approximately 0.5 per cent magnesium oxide to the feed powder and annealing the boule in oxygen.

The various diamond testers would reject rutile but in the absence of the appropriate instruments (including a lens) the amateur could be deceived, since rutile can be most spectacular.

The calcium molybdate *powellite* has been doped with rare earths to give pink, perhaps from holmium (this colour has been reported but no doubt other colours can be made). A variety of fluorescent effects are achievable so despite the low hardness of 3.5, stones are cut from crystals intended for applications other than ornament. The RI is 1.924–1.984 and the SG 4.34.

Wulfenite is the lead analogue of powellite; it is lead molybdate. Colourless synthetic crystals have been grown but while they do not show the magnificent orange of natural wulfenite, they do have a high dispersion. Zircon crystals, purple from vanadium doping, are beautiful but rare.

Proustite is a magnificent, transparent deep red but soft and liable to surface alteration if exposed to light for prolonged periods. The properties of the synthetic material are the same as for the natural. The hardness is 2.5, the RI 2.79–3.08, the birefringence 0.296 and the SG 5.57–5.64. The faceted synthetic proustite specimens I have seen are quite large compared with almost all faceted natural specimens. However, should a faceted proustite turn up at a show there will be a great deal of hype surrounding it, and the true nature of the specimen will not be easy to establish.

In the late 1960s and early 1970s minerals of the *spinel* group were being grown for a variety of research purposes. Many were grown by the flux-melt method and some crystals, though not in great numbers, must have reached the mineral market. A variety of compositions and colours were grown and small crystal groups achieved. These groups will hardly ever get near the market and there may be very few of them in any case. In a collection made by one university some vanadium-doped crystals were blue, some crystals doped by copper were an attractive green, some doped by manganese were red. All were well shaped octahedra and grown by the flux-melt process.

The mineral *gahnite*, also a member of the spinel group but containing zinc instead of magnesium, has been grown to give blue octahedra with an RI of 1.805 and an SG of 4.40. The hardness is 7.5–8. It is possible that some crystals may have been faceted.

A synthetic blue *forsterite* coloured by cobalt gave an RI of 1.635–1.670, a birefringence of 0.010 and an SG of 3.26. Very small gas bubbles could be seen and there was a strong pleochroism giving violet, blue and purple. The high birefringence distinguished the stone from tanzanite, which it was presumably intended to imitate.

Synthetic forsterite grown by crystal pulling can be doped to give a variety of colours but the green is not very like that of peridot, although the properties overlap. The SG is lower than that of natural peridot.

A zinc sulphide with the same composition as *sphalerite* has been grown with a single RI of 2.30 and an SG of 4.06. The hardness is 3.5–4.0.

A borosilicate glass with the name *Laserblue* owes its colour to copper as a dopant. The colour is an intense medium dark blue and specimens give an RI of 1.52. The hardness is 6.5 but the material is reported as heat-sensitive and thus difficult to cut. Another 'named' glass is *alexandrium*, a lithium aluminium silicate glass doped with the rare earth neodymium to give a light blue to lavender range of colours. This glass, with a hardness of 6.5 is heat-sensitive and has an RI of 1.58. The neodymium will give the usual fine line absorption spectrum, a test which will immediately rule out any natural material.

The name *Bananas* (derived from barium sodium niobate) has been given to a yellowish, highly dispersive material with a single RI of 2.31.

Perovskite is calcium titanate and has been synthesized to give a number of colours, of which I have so far seen only pink. The hardness is 5–6, the mean RI 2.40 and the SG 4.05. Colours are achieved by doping and no doubt some will show rare earth absorption spectra.

The tin oxide *cassiterite* is reported to have been grown, but I have seen no specimens. The crystals are said to be colourless to slightly yellow, with an RI of 1.997–2.093, a birefringence of 0.098, a dispersion of 0.071 and an SG of 6.8–7.1. The hardness is 6–7.

Synthetic *fresnoite* was described in the April 2000 issue of *Australian Gemmologist*. The specimen was a 6.11 ct faceted yellow stone. The RI was measured at 1.765–1.770, birefringence 0.019, SG 4.45, hardness 3–4. Specimens showed rounded gas bubbles. The material was the synthetic version of a barium titanium silicate found in California, but in sizes too small for ornamental use.

APPENDIX I

Gemmological Associations

Gemmological bodies, in most countries where they exist, test gemstones for the trade and run courses for students. Most also produce journals.

The Gemmological Association of Great Britain (Gem-A) is the oldest gem testing/teaching body in the world. The *Journal of Gemmology* has high scientific standing and publishes many ground-breaking refereed papers. It is published quarterly. The Association can be reached at 27 Greville St, London, EC1N 8TN. The e-mail address is information@gem-a.info and the web site is www.gem-a.info.

The Gemological Institute of America also tests, examines and publishes and is based at 5345 Armada Drive, Carlsbad CA 92009. The web site is www.gia.edu.

The German equivalent, publishing the quarterly journal *Gemmologie*, is at P.O. Box 12 22 60, D-55714 Idr-Oberstein, Germany. The website is www.dsef.de and the e-mail address is info@gemcertificate.com.

The Association Française de Gemmologie is at 7 rue Cadet, 75009 Paris, France. The email address is gemmes7@wanadoo.fr.

APPENDIX II

Trade Names

The multiplicity of trade names given in the previous (1983) edition of this book seems to have diminished; many of them are never used and few, if any, have been added. New ones arise from time to time and some of them can be found in the text, although they seem of little importance. Since some trade names appear in scientific as well as commercial texts, most of the longer-established examples can be found below. Some names derive, through non-English languages, from the names of chemical elements and others are clearly misapprehensions of the nature of the material.

Cubic Zirconia

CZ
Cerene
Cubic Z
Cubic zirconia II
Cubic zirconium
Cubic zirconium oxide or dioxide
Diamon-Z
Diamond-QU
Diamonair
Diamonair II
Diamonesque
Diamonique III
Diamonite or Diamondite

Diconia
Djevalite
Fianite
Phianite or Phyanite
Shelby
Singh Kohinoor
Zirconia
Zirconium
Zirconium yttrium oxide

Synthetic Corundum

Diamondite
Crown Jewels
Walderite
Violite

Glass

Royalite

Synthetic Rutile

Astryl
Brilliante
Diamothyst
Gava Gem
Jarra Gem
Johannes Gem
Kenya Gem
Kimberlite Gem
Lusterlite
Miridis
Rainbow Diamond
Rainbow Gem

Rainbow Magic Diamond
Rutile
Sapphirized titania
Star-Titania
Tania-59
Tirum Gem
Titangem
Titania
Titania Brillante
Titania Midnight Stone
Titanium
Titanium Rutile
Titanstone
Zaba Gem

Synthetic Spinel

Alumag
Corundolite
Erinide
Magalux
Rozirdon
Strongite

Strontium Titanate

Bal de Feu
Diagem
Diamontina
Dynagem
Fabulite
Jewelite
Kenneth Lane Jewel
Lustigem
Marvelite
Pauline Trigere

Rossini Jewel
Sorella
Symant
Wellington
Zeathite
Zenithite

YAG

Alexite
Amatite
Astrilite
Circolite
Dia-Bud
Diamite
Diamogen
Dimonair
Diamone
Diamonique
Diamonite
Diamondite
Diamonte
Di'Yag
Geminair
Kimberly
Linde Simulated Diamond
Nier-Gem
Regalair
Replique
Somerset
Triamond
Yttrogarnet

Some of these names, with many other, older ones, appear in Peter Bayliss, *Glossary of Obsolete Mineral Names*. Some others may be found in Andrew Clark, *Hey's Mineral Index*.

APPENDIX III

Libraries

Libraries with in-depth scientific collections are not always easily accessible. In the UK the largest and most centrally situated is the British Library at St Pancras, London. This is the country's largest scientific library by far and also holds the patents of all the countries that produce them – they are very useful if you are looking into crystal growth.

To use the British Library collections a reader's ticket is essential. They are issued only to those who need to consult items that cannot easily be found elsewhere and this rule is strictly enforced. Similar rules are in force at all major libraries; the other really large ones in the UK are Cambridge University Library and the Radcliffe Science Library, Oxford. Both are open only to members of the respective universities and hopeful readers from elsewhere will have to make out a good case for admission in writing – never just turn up.

Fortunately, however, there is the British Library Lending Division (BLLD), to which all public and university libraries have access. It has a very comprehensive collection on all subjects and any library will be able to search its holdings for you. You can reach them on the internet yourself.

The four libraries have at least some of their catalogues on the internet. The British Library catalogues, including BLLD, can be found on www.bl.uk/catalogues.

For the libraries of Oxford University, check www.lib.ox.ac.uk/olis, and for those of Cambridge University www.lib.cam.ac.uk/library/catalogues.

Bibliography

BOOKS

Crystal Growth

A knowledge of the principles and practice of crystal growth is essential for the producer of most kinds of artificial gemstones and those wishing to identify the products must have some awareness of how specimens are grown. This means that the extensive literature of crystal growth must be researched. This literature is too extensive to be recorded in full here, but the following entries provide a selection.

Arem, J.E., *Man-Made Crystals* (Smithsonian Institution Press, Washington, 1973)

Brice, J.C., *Crystal Growth Processes* (Blackie, 1986)

Gilman, J.J., *The Art and Science of Growing Crystals* (Wiley, New York, 1963)

Handbook of Crystal Growth (Elsevier Science BV, Amsterdam, 1994)

Keesee, A.M., *Crystal Growth Bibliography*. 3 vols. (IFI/Plenum, New York, 1971–83)

O'Donoghue, M., *Synthetic Gem Materials* (Worshipful Company of Goldsmiths, 1976)

O'Donoghue, M., *Crystal Growth: A Guide to the Literature* (British Library, 1988)

Pamplin, B. (ed.), *Crystal Growth*, 2nd edn (Pergamon Press, 1980)

Wilke, K.T., *Kristallzüchtung* (VEB Deutscher Verlag der Wissenschaften, Berlin, 1973)

Gemstone Enhancement

Nassau, K., *Gemstone Enhancement*, 2nd edn (Butterworth-Heinemann, 1994)

Themelis, T., *Beryllium-treated Rubies and Sapphires* (Available from the author at www.themelis.com)

Themelis, T., *The Heat Treatment of Ruby and Sapphire* (Available from the author as above)

Themelis, T., *Flux-Enhanced Rubies and Sapphires* (Available from the author as above)

Mineral Names and Species

Bayliss, Peter, *Glossary of Obsolete Mineral Names* (The Mineralogical Record, Tucson, Arizona, 2000)

Clark, Andrew, *Hey's Mineral Index* (Chapman & Hall for the Natural History Museum, 1993)

Embrey, Peter and Fuller, John, *A Manual of New Mineral Names* (British Museum (Natural History), 1980)

Hey, M.H., *An Index of Mineral Species and Varieties Arranged Chemically*, reprinted with corrections (British Museum (Natural History), 1975)

Patents

MacInnes, D., *Synthetic Gem and Allied Crystal Manufacture* (Noyes Data Corporation, Park Ridge, N.J., 1973). Covers United States patents only.

Yaverbaum, L.H., *Synthetic Gems Production Techniques.* (Noyes Data Corporation, Park Ridge, N.J., 1980). Continues and updates MacInnes above.

Synthetic Gemstone Production and Identification

Anderson, B.W., *Gem Testing*, 10th edn (Butterworth-Heinemann, 1990)

Barnard, Amanda, *The Diamond Formula* (Oxford, 2000)

Campbell Pedersen, Maggie, *Gem and Ornamental Materials of Organic Origin*

Elwell, D., *Man-Made Gemstones* (Horwood, 1979)

Michel, H., *Die Künstlichen Edelsteine*, 2nd edn (Wilhelm Diebener, Leipzig, 1926)

Nassau, K., *Gems Made by Man* (Chilton Book Company, Radnor, PA, 1980)

O'Donoghue, M., *Synthetic, Imitation and Treated Gemstones* (Butterworth-Heinemann, 1997)

O'Donoghue, M., *Identifying Man-Made Gems* (NAG Press, 1983)

O'Donoghue, M. and Joyner, L., *The Identification of Gemstones* (Butterworth-Heinemann, 2003)

JOURNALS

Crystal Growth

Journal of Crystal Growth

Progress in Crystal Growth and Characterization

Synthetic Crystals Newsletter (published in Sevenoaks, by Michael O'Donoghue)

Conference proceedings in related disciplines often contain papers of relevance to ornamental materials testing

Index